`MW00768026`

THE AWAKENING WORD

Robert W. Mitchell

authorHOUSE®

AuthorHouse™
1663 Liberty Drive
Bloomington, IN 47403
www.authorhouse.com
Phone: 1-800-839-8640

First published by AuthorHouse 11/23/2011

ISBN: 978-1-4670-6736-2 (sc)
ISBN: 978-1-4670-6735-5 (hc)
ISBN: 978-1-4670-6734-8 (e)

Library of Congress Control Number: 2011919125

Printed in the United States of America

In Loving Memory of my Mother
Margaret W. Alexander
10-21-1923 - 11-20-2002
Who Gave me a Childlike Faith

Acknowledgments

I would like to acknowledge the support of the former members and board of the Global Heart Dallas, CSL, as well as others in my life who have shown me such generous love and support. Through the process of writing this book, I have found that people who give so freely of their time, talent and treasure seem to understand that we are all in this thing call "life" together, and together we can make wonderful things happen. This book is one of those wonderful things.

Thank you.

Table of Contents

"Because we are destined to love,
It's time to stop standing alone."

Introduction

In June of 1996, I walked into a bookstore in Dallas, Texas and stood in the middle of an aisle and asked God to change my life. I reached out and picked up a very large book called the **Science of Mind** by **Ernest Holmes**. I opened the book and read part of a short paragraph, flipped over a few more pages, and read a little more. I then went to the checkout and bought it. I did not pick up another book that day to make sure that that was the right one; I did not have to. Something in me knew that God had answered my plea and knew that my life would change that day. And it did.

At that time in my life I was not unhappy, but I was not happy. I was not ill or broke or lonely. I had a successful career and a nice house. On the outside everything looked great, but on the inside something was missing. It was God. I was 51 years old and had always believed that there was "something" in the universe, a Creator. After all, something had to get us here, and I even called it God. But I thought God was for other people, and if I did not bother him, he would not bother me. Funny thing about believing things like that, the more that we see God as being out there the more it seems to be true. So for me, God was something – but not in my life. Later I would realize that all of my life had led me to that one moment when I reached for the book. That moment was a turning point. I was ready for God in my life.

Growing up, the adults in my family were not church goers. My uncle was a Mason and my aunt an Eastern Star member, but they were not active. Although I never went to church as a child, my aunt did take me to a sunrise service in a large open field in East Texas one Easter Sunday. That day I saw how moved she was by her beliefs. My family had to have believed in something. My mother was a single parent with two little boys, so we lived in a world with very little, yet we always had enough. I remember many times when things happened that seemed almost miraculous. Now I look back and know that we had faith, call it life or God, but faith in something.

I also remember as a 12 year old boy being mesmerized by Moses in the movie **The Ten Commandments**. I saw the movie more than 30 times in one month. I know the time frame because I had started skipping school a couple of weeks before the movie came out. Reason? There are many. I was not doing well. I had problems with teachers. So one day I just did not go, and that day led to another and another. How did I get away with it? At the time my mother, my brother and I had moved in with my aunt and uncle (again), and the school did not have their phone number or address, and they had the wrong phone number for my mother's work. My mother never really asked about homework or how things were going at school. No one in my family knew that I was not going to school. The problem was what to do with myself. I could not just stay home because my uncle was there most of the time. So each morning I had to leave home like I was going to school, but stay off the streets because of Truant Officers. So the movie was perfect. Sometimes I would hide behind seats in the back of the theater so that I could see it a second time in a day without paying again. I do not remember how much the tickets cost, but between collecting money from my paper route, pocket change from my aunt and uncle, and being very creative I was able to see the movie as often as I wanted until I was stopped by the police. That changed everything. But for a time, the theater and the movie offered me a sanctuary, a hiding place from school and the world and, consciously or unconsciously, gave me something to think about for the rest of my life, "God."

The Science of Mind gave me a new understanding of God and reintroduced me to the idea of God's Laws, but not laws as commandments, like the movie, but rather Universal Principles as Laws that we use every moment of our lives. Like everyone else, I was already using these Principles and had been all along, but I was not conscious of it. Learning how to use these Laws frees us from the bondage in our lives and can create wonderful things for us. I do not think there is a person alive over a certain age who has not heard the words "cause and effect," but how does that work?

The day I bought the book I began a path to find out about these Laws, even though I did not know to call it a path. I just knew that somehow I was different, the world was different, and that book held the key. That day even sparked the idea that I would become a minister and tell people about God's Laws, although I did not think about how or when that would happen. Nor did I know that the book was connected to a church. I did not read inside the jacket cover of the book where it talks about Ernest Holmes founding the world-wide movement **Religious Science**. For me it was just a book, but a powerful book about a man's ideas about God, life and the universe.

I spent a year reading the whole book. And it was a wonderful, eye-opening year. It was a cleansing for me. I cried a lot, because I was given many answers about my past and my family and how we had dealt with life. I found I was able to let go and forgive. I found out wonderful things about God, the universe, and myself. The most central thing I found was that when I think, according to the emotion and conviction of my thought, something happens, and the experiences of my life are created. If I am to be happy, I must think happy. To be loved, I must think love. To be prosperous, healthy – all the wonderful things in life – I must think them; after all, thoughts are things. Even more, I found that God is for me and not against me, as I had believed before, and yes, my thinking had everything to do with it.

Although I have stepped out in many directions and taken many different paths, my central path is still the Science of Mind. **Dr. Ernest Holmes** tells us, **"The Science of Mind is the study of Life and the nature of the laws of thought; the conception that we live in a spiritual Universe; that God is in, through, around and for us."** This is the path I started in 1996. In 1997 I found my first Religious Science Church now called CSL Dallas (Center for Spiritual Living) under the direction of Rev. Dr. Petra Weldes, to whom I extend my deepest gratitude for the spiritual teaching she has given me, her church, our community, and the world. There is a wise old saying, "If the student is ready, the teacher will appear." I found this to be very true. Petra, thank you.

My spiritual education has been supported by wonderful books, learned teachers, and the paths that I have taken and continue to take, while on my journey. I believe the "**wake up**" process has a lot to do with what we do with our spiritual education. I am reminded of the concept; faith without works is dead. After we find out wonderful things about ourselves, God and the universe, we have to do something about it. Paths are all about doing. In this book you will find many encouraging words about different paths and the rewards and challenges that come with them. You will find that although I do write about my path, the path of **New Thought**; I do not write about the major paths of Christianity, Hinduism, Buddhism, or Judaism. Instead I chose to write about those paths that we can be drawn to for an hour, a day, a month or a lifetime that are in service to our humanity and Spirit. I also do not write about the paths of illness, death, or the dark night of the soul. Although we all take these paths at one time or another, I will save those for a different work.

You will find that paths do not have to be organized or supervised, but they must be approached with consciousness and intent and contribute to our lives, and the life of the world. This book is one of my many paths within my one journey. It started as my blog, "**The Awaking Word**," where I post essays, poems and prayers. The blog grew into this book. My goal here is to, in some way, touch those who have not yet found a path or those who may be looking but do not know what they are looking for – those who may be wondering about life and God as I was, and still do. If you are at a place and time in your life where you are wondering or feel as though something is missing, this book may be for you. It is written for the "beginners mind." The short easy to read essays, poems and inspiring quotes are meant to send your mind off into wonderful directions, and your life on amazing paths that allow you to grow and blossom, and be the divine person you were meant to be.

The different chapters address life and how we live it; our humanness as well as our divine nature. Some of the essays are about the rules of life, life as a game or a gamble, getting ready for our paths and stepping on to them. Although I wrote the poems, it is hard for me to take credit because they were literally given to me during the nights and early morning hours while writing the essays. Like American novelist **Henry Miller, "...it was all given, straight from the celestial recording room ..."** I put my thoughts about that in the poem, **"The Angels of Poets are Upon Me."**

The wonderful art work of the fish on the cover and inside the book is by artist **Diane Stewart**. The quote at the beginning of each chapter is from the poem that will end of each chapter. Diane and I chose fish because in one way we are like fish. Fish swim in an ocean of water and ask **"What is water?"** We swim in an ocean of God and ask **"What is God?"**

In the long run, you will find that our many paths within our one journey are ultimately about love. When we wake up to the truth of who we are, we will find love. Love is the ultimate impulsion in the universe, and you and I are a product of that impulsion. Waking up to who we are and stepping on to our paths begin with the recognition that within our hearts is the Heart of God. Even though we have many differences, there is a sameness within us all, and that sameness is God. At some point that sameness has to come together. That is the journey we all share. We are destined to love, and if we allow Love to point the way, Law will make the way possible.

Reverend Robert W. Mitchell

"The spiritual path is like climbing up to the mountain top through hills and dales and thorny woods and along steep and dangerous precipices. If there is one thing which is most necessary for a safe and sure arrival at the top, it is love. All other qualities which are essential for the aspirants of the Highest can and must come to them if they faithfully follow the whispers of the unerring guide of love."

Meher Baba

THAT CRAZY MOON

If we pressed our hearts together,
The sun would melt like butter,
And beams of light would flow into darkness,
Everywhere.

If we held hands,
The world of separation would fall silent.
Those old stories would stop telling themselves all
Together.

If we danced,
The streets would be filled with flowers.
The music from the spheres would play on
Forever.

If we kissed,
The sky would burst into light a thousand times.
Stars racing across a sea of blue,
How fantastic.

Because we are destined to love,
It's time to stop standing alone.
And so, here we are
You and I.

Now, if that crazy moon would show us its rings,
I would buy one for you...

"Wake up, wake up,
I've called you a thousand times.
You've been crowned Lord of the land,
The sky is your playground;
the earth your slave."

Chapter One:
One Journey, Many Paths

THE JOURNEY & THE PATHS

Journey or path, which one are we on? The answer is both. We are all on a journey, even though a lot people do not know it. And we take all sorts of paths or side trips during the course of our lives. So for each of us, there can be many paths but only one journey. That journey is the journey of self-discovery. **Herman Hesse** tells us that, **"Each man's life represents a road toward himself."** The journey is a lifelong journey; there are paths, however, that we can take for a reason, a season or a lifetime. We can take a particular path that may last a lifetime, but, at the same time we are still on the larger journey. For those who are conscious of both, there is a gift. That gift is the awareness that life is more than its experiences.

We are called to certain paths in different ways. We can be drawn by what we want to experience in our lives or what we would like to see the world experience. If we want to live in a more compassionate world, we could walk the Path of Compassion, or the Path of Christ, or the Path of Charity or Service. If we want to experience inspiration or peace, we walk those paths. However, while taking any path, we need to remember two things. First, true happiness is a result of the journey and not of a particular focus. Being truly happy is a lifelong body of work. Second, you must always remember that whatever path you take, it is you and you alone who take it. You cannot take the steps for others and they cannot take them for you.

Gandhi said that we have to be what we want to see in the world.

So the path to any change we would like to see begins with us. That can be a challenge because sometimes we do not believe enough in ourselves or stay on any path long enough to discover or fully embrace what is there for us. We can give up too soon. We also can jump from one path to another. However, if we allow it, our paths can guide us to a better way of living and an understanding of who we are. They deepen the experience of the journey. And we can have a lot of fun along the way.

Our paths are all about what can be done; our journey is all about who is doing it. Our paths open the doors in the walls we have falsely built in consciousness that keep us from stepping into the reality of God and the fullness of life. The paths are about the love, the service, compassion, peace and wisdom in the world. The journey is about the Self-Realization of our human and divine nature. If we look at the big picture we will ultimately see that we all share One Journey, the journey of the awakening of humanity, the many paths are the different roads we take along the way. I trust that this book will inspire you to step onto a path that will enhance your journey.

"If life's journey be endless where is its goal?
The answer is it is everywhere.
We are in a palace which has no end,
but which we have reached.
By exploring it and extending our
relationship with it
we are ever making it
more and more our own"

Rabindranath Tagore

THE PATH

Although there are many paths that one can take in the world, this book is about spiritual paths. The spiritual path can be specialized and be called the path of Christianity, or New Thought, or peace, or the path of love, or wisdom. Some are organized, some are not. Some are groups and some are not. Some of my paths have been: singing bowls, Taize' services, prayer beads, meditation circles, becoming a Practitioner, then minister and many, many others. The months leading up to my ordination was a path. I have also done things in the secular world that might not seem like a spiritual path, such as attending college. But I find being consciously aware constitutes what I call a path. Anything we do with the conscious awareness that we are both human and divine serves self-awareness. We know that prayer and meditation are paths, but what about standing in line at the grocery store? I find that we can have the spiritual experience anywhere, because God is everywhere.

The Bible speaks about the path or pathway in many places. We are told that if we take a spiritual path it will lead us to our higher purpose and that we will find fulfillment, a state of bliss, love and inner peace. Mostly, it seems as though the path we step on is to find or realize something that at first might seem outside ourselves, when in reality, it is not. **Rumi** says it best – **"You wander from room to room hunting for the diamond necklace that is already around your neck."** We look for God in a book or an exercise or a course of beliefs set down by others. And what we eventually find is all of these things only point the way. I did not find God in the **Science of Mind**, but I did receive a new understanding of God that helped me believe and gave me direction. As a minister, I too want seekers to follow our path and take our courses and believe what we teach. However, I believe that each life is called to the path or paths that will serve it best. We are guided to the experiences and conditions that will help us. Our job is to recognize it and make the best of it. There was a reason you chose this book or the universe gave it to you. Maybe you are ready to begin the "wake-up" process, and this is just part of the process.

But wherever we are, or whatever path we are on, we should always remember that we are like Dorothy and the other characters in the Wizard of Oz; we have everything we are looking for. Our paths take us home where we wake-up to the truth of who we are.

So, are you on a path? If not, are you being called to a path? Consider the following questions:

+ Do you have synchronistic (unexplainable) experiences?

+ Do you find yourself daydreaming?

+ Are your sleeping dreams more vivid?

+ Are you drawn to ask questions and have an overpowering desire to know?

+ Do you sometimes feel an uncommon sense of love or compassion?

+ Do you sometimes cry or get emotional for no reason?

+ Do you think about changing the world?

+ Do you find yourself being the observer?

+ Do you think about God?

+ Do you think about prayer and or meditation even though you may not do anything about them?

If you answered yes to as little as three of these ten questions, you are being called. If you know that you are being called or have unconsciously started a path, you can start making conscious choices and choose your own course of actions and what you want to experience. Knowing is taking control (the good kind of control) of the process.

"There are many paths to enlightenment. Be sure to take one with a heart."

Lao Tzu

WAKE UP, WAKE UP, I'VE CALLED YOU A THOUSAND TIMES

Wake up, wake up, I've called you a thousand times.
In your sleep the moon has wandered and the stars fell,
The night birds have flown away and now the Sun is high.
Once again, you missed the dawn and the noonday bell.

Wake up, wake up, I've called you a thousand times.
There are sugary cakes and fresh water from the well,
The market has fat fish and the best cheeses and wine.
Why sleep when they have so many good things to sell?

Wake up, wake up, I've called you a thousand times.
You're missing the best part of the day,
The streets are filled with beggars.
Wake up, grab a nickel before they go away.

Wake up, wake up, I've called you a thousand times.
The trees are full of apples, blossoms as big as the sun,
The cobbler won't be the same without you my friend,
Still you sleep, not knowing your life has begun.

Wake up, wake up, I've called you a thousand times.
The wind is in our favor, the clouds dance and play,
The air is sweet, the grass is great for rolling,
Yet you sleep this day, like every other day.

Wake up, wake up, I've called you a thousand times.
You've been crowned Lord of the land,
The sky is your playground; the earth your slave,
But asleep, you don't know what you command.

Wake up, wake up, I've called you a thousand times.
But it seems a powerful dream keeps you from me,
A dream that keeps you searching for love and joy.
Wake up, wake up, they're here awake in me.

"I heard God was giving out buds today;
The white ones are especially rare.
But all colors hold a dream to dream,
So take one, if you dare."

Chapter Two:
Are You Ready?

ARE YOU READY TO BLOSSOM?

"And the day came when the risk to remain tight in a bud was more painful than the risk it took to blossom."

Anaïs Nin

When I think about the paths that I have taken, I think about the growth that I have experienced. When I think about Anaïs Nin's quote, I think about how painful life can be without growth. Although I love to look at buds, the bloom is the flower's full potential. The idea of **"Blossom"** in relation to our lives, for me, is all about being open and mature enough to evolve into our full potential. The natural course for a flower is to bloom, which is growth, and growth is a natural state for us too, although growth means change, and change can be scary. The idea of **risk** in the quote is all about our fear of change. **Marie De France** said, "You [we] have to endure what you [we] cannot change." But we do have the power to change. We can change ourselves and our circumstances, but often worry about what is next. The blooming of the flower represents liberation from the bud. Our liberation comes from freeing ourselves from our present state in order to become what we can be. Liberation for us comes from the realization of our potential; just as the seed of the flower contains its full potential, so does the seed of life within us. We do not have to look outside ourselves or to another to blossom. We do not have to become someone else to blossom. We do not have to have permission to blossom. We do not have to be in someone else's garden (theology) to blossom. It is within us to do what God and nature have given us to do. **BLOSSOM.**

"One can indeed help the being to grow...but even so, the growth must come from within."

Sri Aurobindo

CAN WE CHOOSE TO BLOOM?

Not every flower blooms; some die as a bud. Did the flower have a choice? We will never know until we become the flower. But we can ask the question of ourselves, can we choose to bloom? I think we can. I think we can because experiences create opportunities for choices. Our own life can take us to a place where it has to move forward, or wither and die. **Flora Whittemore** said, **"The doors we open and close each day decide the lives we live."** We can close the door on what is and open the door to what will be through the choices we make. Ask any addict about the path of addiction and doors opening and closing through choices. A recovered addict can tell you about the moment of change. Sadly that moment usually comes at a dark time, but if they are willing, that dark moment can be the greatest moment.

Struggle creates change. Our struggles create a desire for things to be different and create change just as it does in nature. For instance, the struggle a caterpillar goes through to become a butterfly - those last few moments, full of panic, fluttering so hard to be released, or the physical pressure on a piece of carbon to become a diamond. If we look at the transformation of life after a downpour or the renewal of a forest after a fire, we can see the change has happened for the good. If we look at the bondage in our lives, our thunderstorms, our fires, we can see that they create a longing for freedom and impel us to transform from one stage of life to the next. We will never really know what choice nature has, but we can choose to stay in the struggle, or we can choose to do something about it. **Anaïs Nin** says that it is a kind of death to elect a state and remain in it. We also can trade one struggle for another or bury our heads in the sand; but in that moment, life is asking us to move on, to change, and if we do we are free, transformed. If you are struggling, good, this is the moment. Maybe it is the moment for you to step out on some wonderful path and free yourself. **BLOSSOM!**

How Do I Change?

"If I feel depressed I will sing.

If I feel sad I will laugh.

If I feel ill I will double my labor.

If I feel fear I will plunge ahead.

If I feel inferior I will wear new garments.

If I feel uncertain I will raise my voice.

If I feel poverty I will think of wealth to come.

If I feel incompetent I will think of past success.

If I feel insignificant I will remember my goals.

Today I will be the master of my emotions."

Og Mandino

BE ALL THAT YOU CAN BE

The idea of a human blossoming, for the most part, means coming into one's own, or coming of age; someone finding their way in a relationship, family, college or career – that is what blossoming as a human seems to be about. Even the **Army** wants us to **"be all that we can be."** But here I am talking about *"blooming spiritually."* Coming into our own as Spirit in our human form. We are both human and divine. We have a God nature and a human nature, and the mission is to bring them together. Blooming spiritually is releasing ourselves from the tight bud of the limited idea of life as our human nature sees it, and living in the fullness of our God Nature. We are here to blossom as our true Self, the Higher Self, and the full idea of life as God intended.

The problem with this idea seems to be that we look at the lives of those who have lived their God Nature and believe it is impossible to measure up. Blooming spiritually does not mean that we perform miracles, walk on water, or do things that would start a new religion or cause a Bible to be written about us. Blooming spiritually means taking our life from where we are right now and seeing past the illusion of separation we feel from our families, friends, community, and world. It is having the faith that our needs and the needs of others will be met every day. It is seeing God in the everyday things in life. It is being kind and loving. It is about stepping out of our comfort zones (although they may not be comfortable) and giving this God Self a chance. **Alan Alda** said, **"You have to leave the city of your comfort and go into the wilderness of your intuition. What you'll discover will be wonderful. What you'll discover is yourself."** Carl Sagan adds, **"Somewhere, something wonderful is waiting to be discovered..."** What if that something wonderful is you?

There is something within us that is wonderful, and it will guide us, if we let it. There is something within us that feels oneness and love, and that something is God. Becoming all that we can be in Spirit means being who we are and living up to the God idea of life in every moment as best we can. Spiritual blooming means being one of those flowers in the garden reaching for full potential. Every path we take helps us bloom; every awareness gives us more life. Every time we take a step, we are one step closer to ourselves. **BLOSSOM!**

"There is no enlightenment out side of daily life."

Thich Nhat Hanh

PREPARE YOURSELF

If we want our lives to blossom and live our full potential, we have to get ready and stay ready. We have to grow our lives to a point where we are ready to receive. We live in God's garden and have been planted in the fertile soil (soul) of GOODNESS, and our work is to prepare ourselves. We prepare ourselves by choosing to live each day as though we are already all that we can be, by healing through prayer, meditation, self-discovery and service. We prepare ourselves by finding the light and staying in it. A flower needs the sun to bloom, and likewise we need the light. Some flowers close up at night because the sun goes away, but God's light shines all the time, so we can blossom anytime. We can stay ready by keeping our dreams alive through desire, inspiration, belief, and positive thinking and actions. We stay ready when we keep moving forward with our lives, standing in our truth and letting God work. Above all, we must keep on, keeping on. As long as we persist we will grow. There is one other thing we can do. In the **Bible**, in the book of **Luke** we are told, **"Ask, and it will be given you; Seek, and you will find; Knock, and it will be opened to you."** Have you asked today? **BLOSSOM!**

"The gracious, eternal God permits
the spirit to green and bloom
and to bring forth the most marvelous fruit,
surpassing anything a tongue
can express and a heart conceive."

Johannes Tauler

If We Can Dream it, We Can Do it

If we want to blossom and live the life we dream, we have to be willing to take chances. We have to step off the edge of the known and take a leap into the unknown. We have to believe that we can make our dreams happen. We have to choose each day to make them happen. We have to keep our awareness of them alive. **Gail Devers** tells us, **"Keep your dreams alive. Understand to achieve anything requires faith and belief in yourself, vision, hard work, determination, and dedication. Remember all things are possible for those who believe."** We have to believe that if we reach for the stars, we can touch them. We have to believe that life is good, and it will be. We have to see our dreams not as fantasy, or something far off, but as reality, here and now.

What stops us from believing in and living our dreams is the loss of confidence, our fear of failing, or not measuring up, or being hurt, when in fact the dream is the assurance; *If we can dream it, we can do it.* Our dreams are not given to us by mistake. They are given to us as something to fulfill: a charge, a task, a path with a goal; the dream has something to prove. All of us have dreams, but most of us do not know that we can dream ourselves *awake.* The more we make our dreams about life and what we can do come true, the more we grow what we believe about life and ourselves. Believing wakes us up. Dreams prove that it **IS** done unto us as we believe. Believe in your dreams and not the circumstances keeping you from them. Look for what you want, and you will find it; dream it and it will happen; act it and you will be it. **BLOSSOM!**

"It's the heart afraid of breaking that never learns to dance. It's the dream afraid of waking that never takes the chance. It's the one who won't be taken, who cannot seem to give, and the soul afraid of dyin' that never learns to live."

Amanda McBroom

I Heard God Was Giving Out Buds Today

I heard God was giving out buds today;
The white ones are especially rare.
But all colors hold a dream to dream,
So take one, if you dare.

Each one has a secret, a mission, a task,
And looks for someone just like you.
They look for those who feel the burn
To make love's dream come true.

If you reach for one and take it,
Will you show it to the sun?
Will you keep alive the memory
That it too came from the One?

Will you tell it all your secrets?
Will you trust what comes your way?
Can you open up your life,
And accept this gift today?

And if you accept this bud,
How will you make it bloom?
What will you say to the rose,
While you're in the garden to groom?

Can you see the Sun in every petal,
The moon eclipsed in the bud,
Do you know that God's in every leaf,
Holding tight to the stem of love.

Can you hold it in your mind and soul,
At first light and under the moon?
And sit with it in those lofty hours,
As you chant inside your room?

And if you take this flower
And make the dream come true,
You'll find it is your heart that opens,
Because the bud and the dream are inside you.

"There is a seed
of life expanding,
Here in
God's Waiting Room."

Chapter Three:
Are You Being Moved?

GET READY

Albert Einstein said, "In our endeavor to understand reality we are somewhat like a man trying to understand the mechanism of a closed watch. He sees the face and the moving hands, even hears it ticking, but he has no way of opening the case." When the moment is right the case will open on its own. But we have to be ready for it to happen. We can get ready for the case to open by moving beyond the resistance of our personal mind and opening ourselves to the Divine Mind, and that takes more than books or sermons or rituals. It takes an inner understanding that we are always in the right place at the right time, if we would only slow down and observe ourselves and the things around us. What are we feeling; what are we sensing; What is it that we can be aware of in the moment? Negative things are calling to be healed. Positive things are meant to be grown. In every moment of life, in everything, there is something for us. Our job is to be aware of it, whatever it is.

Opening the case of the watch is like a move in a masterful chess game. It is both mental and physical. Our consciousness and our physical lives can lead us to moments of opportunity. Our lives are moved around until we reach that one square that changes the game as we know it. We are led to do things or compelled to move in certain directions until that one luminous moment that check-mates the littleness of our humanity, and there, our transformation begins. But we have to seize the moment.

My life led me to a book (**<u>Science of Mind</u>**). The book led me to a church (Religious Science). The church gave me my path by giving me a new way of looking at God and the universe and my place in it; my life changed drastically. Our paths can take us to a place where we can be blinded like Saul or able to see in a new way like Juliana of Norwich, or, like John the Baptist, bear witness to the revelation of God on earth. **GET READY!**

*"A cloud does not know why it moves
in just such a direction
and at such a speed, It feels an impulsion...
this is the place to go now.
But the sky knows the reasons
and the patterns behind all clouds,
and you will know, too, when you lift yourself
high enough to see beyond the horizons."*

Richard Bach

SATORI

Satori is a Japanese Buddhist term for enlightenment. The word means *"understanding,"* but translates as a flash of sudden awareness. A flash of sudden awareness...what would that be like? What would it be like if the universe suddenly opened up and took you by surprise? **Meher Baba** said it is like **"A sort of eruption."** **Alan Watts** said, **"It is like the oft-recurring tale of coming upon an unexpected door in a familiar wall, a door that leads into an enchanted garden."** **Richard Bucke** said he found himself **"Wrapped in a flame-colored cloud."** And many others testify to what could be called a burst of reality. Moments of Divine Revelation that penetrate our human identity and melt it into the **"I Am-Ness"** of existence.

Moments of true consciousness are usually fleeting and for the most part seem more a dream than reality. How does this happen in a single moment or happen to us at all? Is it the result of a lifelong body of work, or desire, or some random divine grace simply looking for a place to happen? Perhaps, this illumination like all illumination is a gift; a gift that the Creator wants to give, or in some way has to give. Perhaps Satori is God's way of letting Itself be known and experienced, and perhaps it only seems like a moment because we live in a time-space reality. Perhaps a moment is all we can handle. Perhaps in that moment there is something for us to do, or perhaps it is just the awakening we need to get us through the rest of our lives. **GET READY!**

*"I was sitting in a train when it happened,
crossing the Punjab to Pathankot and
reading a book on Buddhism. And sitting
there in that crowded train, with all its
heat and smell, suddenly it was utterly
self-evident that I did not exist in the
way I thought I did. And this realization
brought with the experience which I
can only describe as a kernel of popcorn
popping. It was as though the inside
came out on the outside and I looked
with wonder and joy at everything..."*

Joanna Macy

RENEWAL

In the physical world the days keep passing by and each one is new. We talk about yesterday as yesterday. But in the mental world our thoughts and beliefs about yesterday can be brought into today as if they belonged to today. We carry the words and actions of yesterday into the current moment and act in the current moment as if it were yesterday. We carry old mind sets into new experiences and repeat the same feeling and actions and therefore, experience the old experience over and over again. We relive old experiences in new places with new people, but it is still the same. We keep getting the same jobs, just in new places; attracting the same relationships, just different names. If we do not renew the mind set we fall into the traps of yesterday. Yet there is a part of us that seeks to make everything new. The Spirit in us seeks to renew our minds. The Bible says through renewal we will have a new heart, a new Spirit, and new experiences.

To renew our mind set we have to empty ourselves of the conditions of yesterday and allow ourselves to be filled with the newness of today. The spiritual solution to the human condition is to let go of who we were yesterday to become who we can be today. Simply put, as the Bible says, be transformed by the renewing of your mind. This is the way we have a new heart, a new Spirit, and a new mind, today. **Be Ye Renewed!**

"If you want to become whole,
let yourself be partial.
If you want to become straight,
let yourself be crooked.
If you want to become full,
let yourself be empty.
If you want to be reborn,
let yourself die.
If you want to be given everything,
give everything up."

The Tao Te Ching

A SEED IS SPINNING THE WORLD

There is a seed of life expanding,
It grows in your heart and mine.
The seed of miraculous existence,
In our sacred flesh entwined.

A seed of light in the darkness,
A fire that burns in the night.
The blessed word of reality,
Singing the song of Light.

Its shell is all that glitters,
It glimmers emerald and pearl.
The poet cries out in tears of joy,
"A seed is spinning the world."

It carries inside it everything,
Everything there is to know.
Every stream and rock and valley,
The flow of the melting snow.

Its hallowed voice of reason,
Rejects uncertainty.
A million moons, a thousand suns,
Are here in its Eternity.

It's the reason for surrender,
To what we are from birth.
The treasure here in this Kingdom,
The face of God on earth.

There is a seed of life expanding,
Here in God's Waiting Room.
Your heart cradles this precious gift,
The seed in the Holy Womb.

"There's a pocket in my soul
I think God is a good tailor to put it there,
A place to keep little things for myself,
And other things to share."

Chapter Four:
Things You'll Need To Give

GIVE LIFE

If you are being called to a path there are things that you need to give along the way. Not just because you might want to give, but because it works in your favor if you do. The Principle of Cause and Effect states that every action has a reaction – everything we put out, we get back. Giving anything gives us more in return, so, giving life gives us more life. By giving life I mean using those qualities that expand and make alive as opposed to those that shrink or diminish.

One of the main tools we have to work with is being positive. Being positive is the highest use of the Law because it expands while negativity shrinks. So love or compassion will strengthen, while anger will weaken and hate will kill. Blame, shame, and guilt diminish life while joy, love, laughter and forgiveness expand life. Having a reverence for all life gives life. Respecting not only those we love but even strangers, gives life.

We give life when we see the world through our God eyes and touch the world through our God actions; such as hugging instead of fighting or smiling instead of frowning. **George Eliot** said, **"Wear a smile and have friends, wear a scowl and have wrinkles."** Hugs and smiles give life, and so does a simple "hello." Taking a moment to acknowledge a stranger gives life and shows that we are not afraid. Fear weakens life, faith gives life and allows us to shine our light; hiding our light darkens life. We are told in **Matthew, "People don't light a lamp and put it under a basket but on a lamp stand ...."** When we let our light shine we are using our ultimate gift from God: freedom. We can use it to give life by living freely, or we can use it to reduce life by living in bondage. God gave us so much, so our holding back only reduces life. Use everything that God has given you. **GIVE LIFE!**

"When I stand before God
at the end of my life,
I would hope that I would not have
a single bit of talent left,
and could say I used everything you gave me."

Erma Bombeck

GIVE LOVE

"Love thy neighbor as thyself"(Mathew 19: 19) is one of the commandments, but how do we do that? Love is a powerful word and one that we have made conditional in our very human lives. We love because of someone's appearance or actions; "**I will love you if ...**" We have love all caught up in romantic feelings and relationships (spouses, parents, siblings, friends) and so it is easy to be confused about what is being asked of us. Our neighbor, that is the world, can be filled with many things and ways of being that we do not love. So the commandment becomes a riddle for us to solve. Each of us instinctively wants to love, but how? How can we take this very personal and powerful thing that we feel and give it to a neighbor that we might not want to love?

To solve the riddle, we can start by looking at love as something that we share with our neighbor. Within us all, within each of us are the very seeds of love given to us at the creation of the universe; therefore, love is universal in terms of motive, as well as personal in terms of meaning. Love is the purest motive force. Love is God in Motion. **Paul** said, "**... love bears all things, endures all things.**" This something within us sees past the transgressions of the world to the presence of God within all beings, and it is only through us that this action happens. Love is something for us not only to have and hold, but to give as well.

Where can we start? We can start by setting aside personalities and actions and focus our love on what God has created. We might look at a grain of sand with love as well as wonderment. Look at a leaf or flower with love as well as amazement. Look at a sunset or sunrise with love as well as awe. Look at a beautiful painting or sculpture with love as well as inspiration. The love that we can feel, or at least imagine, is God in everything. Love in its spiritual sense is the Idea and Dream of God that lies within creation. Love is what we were conceived in and our being longs to experience and give it.

We can look at the riddle in another way. We actually do keep the commandment, but maybe not in a conscious way. In many ways we do love our neighbor, _as ourselves_, and that is the problem. We may not love ourselves very much, and we cannot give away something we do not have. We are all a work in process, and whether we perceive ourselves as perfect or in need of change, we can love ourselves just the way we are. Learning to love ourselves is the first step to seeing the love in the world that we want to see. So we have to give love to ourselves. Then we are able to give it to our neighbor. **GIVE LOVE!**

"When we let the love that is within us go out to the God who is in all people and the Divine Presence that is in all things, then we are loving God with all our heart and with all our soul and mind because we are recognizing that the Spirit within us is the same Spirit that we meet in others. This is loving our neighbors as ourselves."

Ernest Holmes

GIVE JOY

Joy follows love: "The fruit of the Spirit is love, joy..."(Gal. 5:22). St Francis of Assisi teaches us that joy is the result of constant prayer and purity of heart. He made it a point to keep himself in joy of heart by knowing **truth** in his heart. He tells us that joy is the result of faith and gratitude. When we increase our faith and gratitude our joy increases. Although this essay is not about finding joy through sorrow, we should realize the power of joy overcomes sorrow. Here, I want to point out that joy is the effect of knowing the truth of Spirit in our hearts and acting out that truth in the world. Truth is joy. It is also contagious; it helps break down barriers between people and shows that one is confident and happy in life. **Mother Teresa** said, "**... joy is a net of love by which you can catch souls.**" Joy, as everything does, works within the **Law**. The **Buddha** said, "**We are shaped by our thoughts; we become what we think. When the mind is pure, joy follows like a shadow that never leaves.**"

The many paths we take are all about growing in our spiritual nature. As we grow, joy becomes our constant companion. When we are joyous we are better prepared to handle situations and stay more focused and positive. We share our joy, or give joy by doing things for others. Joy is living a life of service, and as we serve others our joy provides assurance and gives hope. Like all things that are lasting and true, joy is found to be a product of the spiritual dimension. To live in joy is to recognize God in all things. The seed of joy is within you! **GIVE JOY!**

*"It is the spirit of a person that hangs
above him like a star in the sky. People
identify him at once, and join with him
until there is formed a paradise of men
and women, thus inspired. No matter
where you find this spirit working, whether
in a person or an entire organization,
you may know that Heaven has dropped
a note of joy into the world!"*

George Matthew Adams

GIVE MEANING

When I think about the word *"meaning"* I always think about the big question, **"What is the meaning of life?"** I believe to answer that question we have to look at the individual meanings we give to our actions in life. Austrian psychiatrist, **Alfred Adler** tells us that **"Meanings are not determined by situations, but we determine ourselves by the meanings we give to situations."** The meaning of life is what we give it. We create meaning; we give meaning to everything, and meaning is what gives depth and character to our actions.

Carl Gustav Jung said, **"The least of things with a meaning is worth more in life than the greatest of things without it."** Even with busy and successful lives so many people feel unfulfilled and empty because of lack of meaning; lives filled with actions that are shallow and vain. Doing something just for the sake of doing it or just to get it over with is worthless and pointless. Without meaning our lives and actions are lost. Meaning fills and makes real the intention of Spirit in our lives. It is the intent that matters. If there is meaning in our actions, Spirit is at work in our lives, and things will have depth and purpose. We have the awesome responsibility to the Spirit's intention and purpose by adding depth and value to our experiences. It is the way we devote ourselves to what we are doing that creates meaning. Are we coming from a place of love, compassion, sincerity, peace or joy? Where we are coming from has the power to create and the power to destroy. The inner importance we give to any action will determine the outcome and what we become in the process.

The meaning we give our experiences is a direct reflection of who we are. **Joseph Campbell** said, **"The meaning of life is whatever you ascribe it to be."** What we ascribe it to be can be found in the meaning we give life. So, the answer to the question, *"What is the meaning of life?"* is ultimately up to us. We can answer: Love, Compassion, Giving, Forgiving, Peace, Joy, and whatever else of real value we choose to add. **GIVE MEANING!**

"For the meaning of life differs from man to man, from day to day and from hour to hour. What matters, therefore, is not the meaning of life in general but rather the specific meaning of a person's life at a given moment."

Viktor E. Frankl

Give God

We can never be God, that is, "GOD." But we are like God, **Gen 1:27**. We can never give all of God or Spirit, but we can give what we have been given. In the **Substantive Properties of Spirit, Thomas Troward** tells us that Spirit is: **"Life, Love, Light, Power, Peace, Beauty, Joy."** We do not see these things as things, but we see and experience their effects. We can see them and experience them because these attributes are in us too. God has given us what God is. We are created in the image and likeness of... and, therefore, have an instinctive longing to give, and to offer what we are.

When we give peace, compassion, forgiveness, kindness, we are giving God. When we offer understanding, truth, wisdom, or meaning, we are giving God. When we open a door for someone or give a dollar to a beggar, we are giving God. When we hold a crying child or spend time with the elderly, we are giving God. We are a little circle within a big circle and every time we give God we widen our circle.

It is our God Nature that causes us to reach out, to care for our planet, animals, other human beings and ourselves. Giving God is how we let God be God in us. The **14th Century Mystic Meister Eckhart** said, **"God expects but one thing of you, and that is that you should come out of yourself in so far as you are a created being and let God be God in you."** For me, coming out of ourselves as created beings means coming out of the idea of separation. That idea alone is what keeps us from being God on earth, the **Christ**, the **Yod**, the **Buddha**, the hand and body of God. Having the nature of God gives us the capacity to be creative, but it also gives us the capacity to give and to care. **Ernest Holmes** says, **"The Gift of God is the Nature of God, the Eternal Givingness. God cannot help making the gift, because GOD IS THE GIFT."** God does not withhold from us, and we should not withhold from the world. The gift from heaven is forever being made, and we should be an open channel for it to flow through us. The Gift of God is found in every path we take. **GIVE GOD!**

"Be grateful for the joy of life. Be glad for the privilege of work. Be thankful for the opportunity to give and serve. Good works is the great character-builder, the sweeter of life, the maker of destiny."

Grenville Kleiser

THERE'S A POCKET IN MY SOUL

There's a pocket in my soul
Where I carry a picture or two,
They're very old and fragile,
But they remind me of what is true.

There's a picture of the Moon
Where I can also see the Sun.
The play of ice and fire together,
Together at last as one.

There's another that looks
Like something very far away,
But if I hold it to the light just right,
I can see angels at play.

There's another that's like a mirror,
It shows the true reflection of me.
When I look at it, I see this little child,
How forgetful can I be?

There's a pocket in my soul
Where I carry bits of paper, mostly simple notes,
Affirmations about Trust and God,
Some are my own, and some are powerful quotes.

There's one about me
"Having a right to be here,"
That I'm nothing less than the trees and the stars,
And that God is always near.

There's another that says
"I have to be what I want to see."
I want to see love – so I look in the pocket,
Love's here too, it's just been waiting on me.

There's a pocket in my soul
I think God is a good tailor to put it there,
A place to keep little things for myself,
And other things to share.

I don't always remember where I keep my truth,
But then my hand gets cold,
And then I'm reminded,
There's a pocket in my soul.

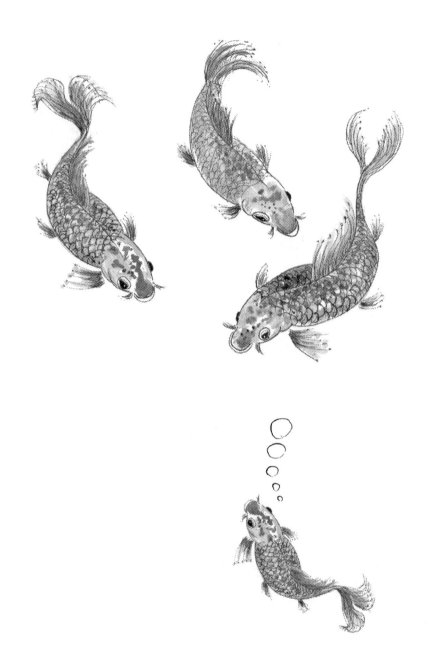

"They tell me about the secrets of love,
The taste of sweetened wine,
Meadows, gazelles, and daffodils,
And that their hearts are mine."

44

Chapter Five:
Things That Will Help

GUIDANCE

If you are stepping out in the world, on different paths doing new and different things, would you like some help? Would you like for someone to guide you and tell you what to do? **Carl Jung said, "There is a deep need in the world just now for guidance - almost any sort of spiritual guidance."** There are, of course, teachers, guides, gurus, coaches, and mystics that teach different paths or ways to help further one's journey to self-realization, and most with real and valid wisdom, intention, and truth. From those who have gone before us and those who are here now, we find proven paths to enlightenment, ways to connect to God, and life's fulfillment. Most religions, spiritualities and philosophies offer directions, courses of action, rituals, and ways of being that can lead us to the secrets of Life. But ultimately, our paths are up to us. We can allow the wisdom and experiences of others to inspire us to move forward with our paths, but like everything else, real guidance, comes from within.

Within every one of us is the Eternal Guide, the teacher, guru, saint, sage, prophet, seer, helper, revealer and counselor, where we find truth in the words in **Psalm 143:8, "Cause me [us] to know the way in which I [we] should walk."** Within us is the Omniscient, Ever Present, all Knowing Self; the Kingdom of God. It is the Divine Guidance within that leads and shows us the way. But the only way it can work for us is by working through us. It works through us or guides us through **meditation, intuition** and **Law.** Allowing Spirit to guide us is a matter of faith, knowing that at each moment we are in the right place at the right time, and at that moment, and if we are open to it, there is something for us to do, or say, or see. In each moment, we are guided to where we are supposed to be. In this way, we can say with **Emerson, "The fact that I am here certainly shows me that the soul had need of an organ here."**

"We believe that the Mind of God governs all things. We believe that God's Intelligence in the midst of us is governing and guiding, counseling and advising, causing us to know what to do under every circumstance and at all times, if we will but trust It."

Ernest Holmes

Inner Guidance

If we go back to chapter two and ask ourselves, "What will help us to bloom?" **Osho** says, **"Meditation... is going to help you flower."** There are many forms of meditation and well documented results of the effects and benefits of meditation, and for most serious seekers, meditation is the backbone of the spiritual path. In this short discourse I cannot discuss all the benefits of meditation, but I do want to address meditation as **"guidance."**

The **Buddha** said, **"Meditation brings wisdom; lack of mediation leaves ignorance. Know well what leads you forward and what holds you back, and choose the path that leads to wisdom."** I do not know who said it, but the words, *"If you do not go within you go without,"* has always stuck in my mind. I do not want to go without, so I go within. I want internal guidance, and like others, I believe meditation is where I get it. Prayer is talking to God. Meditation is listening to God, although it is not like a conversation we have with people. In meditation we listen, but not with the ears on the sides of our heads. We listen, not with sound, but Spirit. And, like prayer, we try to make meditation complicated. Meditation can be as simple as a few moments in silence, being the observer of one's thoughts without attachment, letting them go without **chasing** after them with opinion, judgment, or question. In this way, we clear the mind to receive.

Our minds are instruments and meditation is a tool that uses the instrument as an inlet for guidance and connection. Meditation is the time spent surrendering the mind to become still enough to listen to that small voice within. That small voice may not speak to us at that moment as something understandable, or audible, or as the voice of Cecil B. DeMille. It may or may not be clear at that moment, but in a day or two, or ten, we may get an idea, an answer, an inspiration, or some direction about something in our lives. Meditation makes it possible for God to guide us.

For more on meditation, see chapter 12.

"My meditation
Is my soul's
Soundless sound-conversation
With my Inner Pilot."

Sri Chinmoy

INTUITION

Intuition is a very abstract concept. A lot of people have misconceptions about what it is. We have all heard people say: *"My intuition was wrong,"* when in fact it was not intuition speaking to them, but only their thoughts or ideas. And we all know those can be wrong. Intuition is God in us, and God simply knows. This is why we say that God is Omniscient, All-Knowing. Our common everyday thoughts tell us that certain things are possible, and they may or may not be. Experience is the only way we know for sure. But intuition works in the realm of Reality. With intuition there are no failures, mistakes or misses. Intuition has immediate access to Truth, and Truth does not have to reason or consider external facts or believe in failure.

Intuition roughly translated means **"to look inside."** It is an insight where we know that we know without knowing how or why. **Ernest Holmes** describes intuition as **"God in man, revealing to him the Realities of Being; and just as instinct guides the animal, so would intuition guide man, if he would allow it to do so."** We allow intuition access by keeping our minds centered and ready by expanding our consciousness through meditation, prayer, study, contemplation and self-observance. Intuition is more than a thought; it is a feeling, an instinct, and it comes from deep within our own nature. We can call it conscience – we can call it a hunch, a vision, or dream; but whatever we call it or however we sense it, it is real and unfailing. **Albert Einstein** said, **"The only real valuable thing is intuition."** Like all things valuable it is uncommon, not because it is unavailable, but because we do not look inside very often.

"Our consciousness expands through meditation and contemplation, through conscious communion with the Invisible, and through our intuition, which is the voice of Spirit in us..."

Ernest Holmes

" **S** plit a piece of wood and you will find me. Lift a stone and I am there."(Gospel of Thomas) This is probably the most haunting phrase that has come to me on my spiritual path, and these words return to me daily for reflection. From them, I understand that God is everywhere; we cannot look at or touch anything without coming in contact with God. God is everything. If we explore deeply with our spiritual senses the most common things in life, we will find **"Life."** Call it waves and particles, call it energy, but within everything, whatever we choose to call it, God is there. **Elizabeth Barrett Browning** said, **"Earth's crammed with heaven, and every common bush afire with God...."** We just do not see the fire most of the time. We take life, our world, and everything in it for granted and do not see the truth behind the statement **"There is but One Life."** It seems only in those moments, when caught up in a heightened awareness of reality that we truly feel this sameness hidden everywhere. Look for the fire. It is everywhere. Find the fire and you'll find the magic of Life.

"There's magic all around us.
In the rocks and trees,
and in the minds of men,
Deep hidden springs of magic.
He who strikes the rock aright,
may find them where he will."

Howard Thurman

THE ANGELS OF POETS ARE UPON ME

Late at night the voices whisper
The ideas that come to my head.
My room becomes a playhouse
As I try to sleep in my bed.

In my dream the voices echo.
I'll awake and look around to see;
I'm alone but the room is crowded,
With angels looking at me.

One by one they make an entrance,
Floating visions of beautiful light.
Golden wings gracefully arching,
For this midnight fantasy flight.

Awake in my sleep I see them,
A gathering of spirits here.
My pillows propped for the hour,
For the moment heaven is near.

The angels of poets are upon me;
The reflections the poets left behind.
Giving life to words and memories,
Recounting their perfect rhyme.

Golden tongues reciting treasures;
Luminous halos begin to spin.
Quills flying like darts in circles,
A gathering I don't want to end.

They speak to me of roses and rivers,
Garden gates and the beloved's door,
A nightingale, a golden key,
The journey we walk evermore.

They tell me about the secrets of love,
The taste of sweetened wine,
Meadows, gazelles, and daffodils,
And that their hearts are mine.

They tell me I must lose myself,
If I really want to be found;
A hollowed reed for the breath to flow,
And humble if I want to be crowned.

The joy of their words alive in me,
Ecstasy swelling my heart.
Meter and rhyme in perfect time,
From angel lips this beautiful art.

I know this student is ready,
Why else would these teachers appear?
My task now pen and paper,
My destiny drawing near.

Is this gathering a reality?
I don't want to wake too soon.
I'm afraid if I turn the light on,
There'll be a hush across the room.

"The Golden Umbilical
Between thought and form,
Giving life to every word and deed."

Chapter Six:
Rules, What Rules?

CAUSE & EFFECT

In Dr. Chérie Carter Scott's inspirational book, **If Life is a Game, These are the Rules,** she points out that our **"...time here on Earth is brief. Time passes and things change. You have options and choices in which to make your wishes, dreams, and goals become reality."** Playing by the rules helps us achieve our dreams and wishes. But like so much of life, playing by the rules is a choice. We can learn what the rules are, but if we choose to ignore them, or break them, like any other game, we will face penalties or fines.

Florence Scovel Shinn says, **"The game of life is the game of boomerangs. Our thoughts, deeds, and words return to us sooner or later, with astounding accuracy."** So breaking the rules can have consequences. **Joseph Brodsky** says, **"Life is a game with many rules but no referee. One learns how to play it more by watching it than by consulting any book...."** Sometimes we watch, and it seems that others are winning, when in fact they are not. We have all seen politicians, movie stars, sports figures, ministers and others rise to fame and then suddenly fall, only to find out they were not playing by the rules.

So, what are the rules? I believe the rules of life are more about Universal Rules and Truths than about individual rules and beliefs. Individual rules and beliefs are personal, and what works for one may not work for another. Universal Rules and Truths are impersonal and work for all and all alike. **The Law of Cause and Effect** as a **rule**, says,

"If you hurt someone, at some point, you will be hurt." At the same time it says, "If you love, you will be loved." It is where the **Golden Rule** of life, **"do unto others…"** comes from and is the first rule of life. We can only expect to get from life what we put into it. It is the first rule of the game and the one that determines the outcome. **PLAY BALL!**

*"Life has often been compared to a game.
We are never told the rules, unfortunately,
nor given any instructions about how to play.
We simply begin at 'Go' and
make our way around the board,
hoping we play it right."*

Dr. Chérie Carter Scott

BE TRUE TO YOURSELF

In the game of life there are many rules: do not drive drunk, do not forget your mother's birthday, pay your taxes, get to work on time; they go on and on. Most of them seem to be somebody else's or society's ideas about right and wrong, or what we should do or not do. We also have rules of our own: brush our teeth before bed, always tell the truth, change the air conditioner filter every 30 days; however, unless someone is watching, if we break our own rules, no one knows it but us.

We can break the rules of life that we have set for ourselves and wind up with self-distrust and self-doubt. The rules of integrity, honesty, faithfulness, love, kindness and **"What is it all about,"** for US rules. We can make them, and we can break them. If we break too many of them too often it can lead to a shift of our self-value and sacrificing our dignity and self-worth. We can lose our integrity in the game, as we know it, and feel as though we have to win at any cost regardless of who we hurt, including ourselves. The cost can be a complete breakdown of who we think we are.

Polonius (Shakespearean Character) said, **"This above all, to thine own self be true, and it must follow, as the night the day, thou canst not then be false to any man."** **Janis Joplin** adds, **"Don't compromise yourself. You are all you've got."** We are the only ones that can honor the rules we set in life and staying true to them is how we stay true to ourselves. Keeping the rules is how we keep the game intact.

Being true to ourselves and what we believe allows us to play the game with impeccability. **Don Miguel Ruiz** says, **"Impeccability means 'without sin.' Religions talk about sin and sinners, but let's understand what it really means to sin. A sin is anything you do which goes against yourself."** Sin is also an archery term meaning missing the mark. When we break the rules we set for ourselves, we miss the mark. Not staying true to ourselves is going against ourselves and the rules we value. The spiritual maxim, **"Be true to yourself"** is the key to the vision we hold of ourselves and, in turn, the vision the world sees. It is the rule that only we can make and break. **BE IMPECCABLE!**

"You cannot believe in God until you believe in yourself."

Swami Vivekananda

WE ARE RESPONSIBLE

We have all heard that **"Life is not fair,"** and it seems that way until we learn the rules and start playing by them. Because we are creative or use the Creative Process, our lives show up as a matter of cause and effect and not the flip of a coin by a god in the sky. We have the coin and, as long as we flip it, we will never know if we are going to get heads or tails, but we can stop flipping it and make choices instead. The coin is our freedom of choice, and we can flip it or we can spend it. The only way we can spend it is to use it.

If we use or spend the coin (do not be afraid to spend it, we can never run out of this currency) it affords us a do-over if we do not like what is going on with the game. I remember playing **Parcheesi** as a child and how much fun it was if I was winning, and, if I was not, there was always the **thrill** of being able to start another game. That thrill for us comes in the understanding that we have the power to start over at any moment, but it is up to us. **Denis Waitley** says, **"A sign of wisdom and maturity is when you come to terms with the realization that your decisions cause your rewards and consequences. You are responsible for your life, and your ultimate success depends on the choices you make."**

Taking responsibility for our choices can be difficult. Accepting responsibility does not mean a guilt trip that says, *"We were bad or stupid or wrong."* It simply means we are ready and willing to do something about it. If we think that we are bound to what is going on, we are. But if we understand the principle that says the future is not bound to the past, we can direct the future here and now through choices. **"Each player must accept the cards that life deals him or her. But once in hand one must decide how to play the cards in order to win the game."(Voltaire)** We can look at life right where we are, right now, and then decide.

Responsibility, when accepted, is a wonderful gift, but we have to let go of any blame, because blame only creates more future reasons for blame. Feelings of blame amount to living in the past and, at the same time, because of the Law of Cause and Effect, creating our future. Blame is holding on to the coin God gave us, squeezing the life out of it, wondering if we should try one more flip. **LET GO!**

"Responsibility is not blame, however, and understanding the difference between the two is crucial to learning this lesson. Blame is associated with fault, whereas responsibility denotes authorship. Blame carries guilt and negative feelings; responsibility brings the relief of not having to dodge the full truth anymore and releases that guilt..."

Dr. Chérie Carter Scott

GROWTH

Growth is a Natural Law in nature, but growth as a rule for life goes beyond the physical. As a rule for life, growth says, "If we do not grow we will never be anymore than we are right now." If we do not grow, our lives can become like soap operas. As **Jane Wagner** says, **"I worry that our lives are like soap operas. We can go for months and not tune in to them, then six months later we look in and the same stuff is still going on."** The reruns of our lives are a result of not growing.

Everything is a process. Life is a process; we are a process; the universe is a process. Did you know the universe is growing? The universe is always in the growth process, expanding a little every moment. For us, growth is a process that is started as easily as making the decision to grow. **Alcoholics Anonymous** has a saying, **"Sick and tired of being sick and tired."** Our paths give us the opportunity to look at the places in our lives where we are not growing and do something about it. God or life is always offering us the opportunity to grow. Our experiences are invitations to grow. How many times do we have to make the same mistakes over and over only to relive them again? **Albert Einstein** said that, **"You can't solve a problem on the level it was created on."** We have to rise above the problem to change it. We have to grow to change the problem. And we have to remember that, life is always changing, but growth is optional. We are always changing, but we do not always grow.

Growing is improving our self-awareness, improving our self-knowledge, building our identity, developing our strengths, and improving the quality of our lives. Growing allows us to realize our dreams. Our growth is not automatic, nor is there any guarantee of individual progress. But if we do the work, there is the guarantee that things will change for the better.

So how do we grow? We grow through knowledge and experience. We grow by getting out of our ruts. We grow by living life, exposing ourselves to new people, places and things. We grow through trial and error and getting out there and doing it. Our chance for growth begins when we learn that we are not our conditions; that we are not

64

controlled by them, and that we can change them. We grow when we learn that we have the **POWER OF CHOICE**. It begins with the realization that we were never meant to be stagnant. We were meant to grow and we can start growing today. **GROW!**

"Every person has the ability to grow. While physical growth may reach its peak, the mental and spiritual growth of each individual can continue forever. Life, as it unfolds from day to day, may repeat many experiences of your past. However, you can never meet a former situation in the same way—provided, of course, that you continue to grow."

Robert Bitzer

LOVE RULES

The goal of the game seems to be simple: have fun and enjoy your time playing. However, if we play as opposed to sitting it out on the sidelines, we have to pay attention to the rules. The rule of **"Cause and Effect"** says, "If you put something out there, it will come back." On the field of life, if we want players to treat us with respect we have to treat them with respect. Playing the game from the **inside out** creates a game with integrity. The rule of **"Responsibility"** changes "Life is not fair" to "If I play fair, the game will be fair." So playing the game responsibly creates respect and harmony. The rule of **"Growth"** says, "If we do not grow we will never be any more than we are right now." If we do not grow we will never pass Go, we will never collect the $200.00, and we will never be able to buy Boardwalk or Park Place.

Like other games, the game of life has a rule that stands out among other rules; a cardinal rule, and that rule is **"Love."** Love as a Rule says, **"You can't play the game without me."** Love breaks all ties, exceeds all boundaries, defines the goals, reverses penalties, calls time, trumps all other rules and directs the course of the game. Love is the checkmate, the final goal and the reason for the game.

Love as a rule in the game of life is the beginning, the end, and the goal of every match, set, inning, and every score. Love is both the reason and the passion behind every play. Love allows us to see God in the other players. Love is the greatest power, the best medicine, and meets every human need, and it is how we play the game well. **Vincent Van Gogh** said, **"Love many things, for therein lies the true strength, and whosoever loves much performs much, and can accomplish much, and what is done in love is done well."** Love encompasses cause and effect, being true to ourselves, responsibility and growth. It is both the rally cry at the beginning of the game and the bonfire at the end. It and it alone sees us all as most valuable players. It defines the game for us regardless of who we are, or who we think we are. It levels the playing field by seeing us as equal. After all, **"At the end of the game, the king and the pawn go back in the same box."**(Italian Proverb) LOVE RULES!

Now the game.
Your game.
The one that only you was meant to play.
The one that was given to you when
you came into this world...
You ready?
Take your stance...
Don't hold nothing back.
Give it everything...

The Legend of Bagger Vance

WHERE IS OUR SACRED SOUL?

Where is our Sacred Soul? It's everywhere.
Hidden in the Forces, silent streams of receptive bands,
Molding the shadows, forming the Presence,
The Cosmic Clay in the Creator's Hands.

The Golden Umbilical between thought and form,
Giving life to every word and deed.
It pierces matter, shaping reality,
It frees the Cosmic Seed.

The Divine Threads of God,
The Blood of the Spirit in us,
The Celestial Cord in the Holy Womb,
Drawing our flesh from out of the dust.

We are all one in this ocean of Soul,
Our bodies only a peak like a wave.
Our individuality seems all too real,
Otherness is Oneness gone astray.

The Soul, the Soul, the Sacred Soul,
The Holy Ghost we cannot see.
The Emanate Force, the Master Loom,
That weaves what we believe.

We gather it like reams of silk,
Making the garments we wear.
The robe of love, the shroud of fear,
The cloak of sadness or despair.

Our soul wants to be set on fire,
To feel the burn of our desire.
Our dreams, our wants, our wishes,
The pleasures our lives require.

The Soul, the Soul, the Sacred Soul,
The Holy Ghost we cannot see.
The Force of Creation, the Hand of God,
Moving through time and Eternity.

"Rambling from door to door,
In rooms bewildered, I walk alone.
Things seem to appear
then disappear,
So nothing seems quite
my own."

Chapter Seven:
Want To Be A Star?

THE ACTOR (PART ONE)

*"All the world's a stage, and all the men
and women merely players: they have
their exits and their entrances; and one
man in his time plays many parts..."*

William Shakespeare

Even though **Shakespeare** was not the first to compare the world to a stage, **"All the world is a stage"** is his most frequently quoted passages. He compares the world to a stage and life to a play, and catalogues the seven stages of a man's life; the infant, schoolboy, lover, soldier, justice, pantaloon, and second childhood. My aim in these next five essays is not to address the seven stages, but the idea of the **actor**, the **play**, and the **stage**. If we are actors in this play of life, we play many roles over the course of our lives. However, we do have an ongoing starring role. That is the character of who we are; our name, identity, and the personality that we grow into. That character stays with us as us. But, in any life there are scenes that require us to play different parts. There are drama scenes, comedy scenes, love scenes, death scenes, scenes that involve others or not, and more. The biggest problem with acting in this play seems to be that we never receive the **script** ahead of time. Our acting is mostly improvisational, responding in the moment to what is going on around us and from our inner feelings. This "on the spot" or "off the cuff" performance can often prove to be bad because we are so unprepared. The scene or moment calls for a proper response that is unknown to us, so we may act badly. Life does not allow rehearsal time, and we seldom get the chance to do things over again. There is no one in the background who shouts "cut." There is no part of life that is just a rehearsal – everything counts.

We play heroes, villains, lovers, friends, winners and losers; both the wise and the fool. We play leading roles, co-starring roles, bit parts, sad parts, and happy parts. Most of us, most of the time, just want a happy ending to whatever tragic or dramatic scene that is playing in the now. The Greek poet **Palladas** said, **"Our life's a stage, a comedy: either learn to play and take it lightly, or bear its troubles patiently."** I think the most important part of Palladas' statement is **"learn to play."** If we look at the great actors of the world we know that they have studied and trained, and that is what the play of life is all about. Life is a process of learning, and if we can see the learning curve in our experiences we are better prepared for the next scene. Rehearsed or not, we are able to respond. The other part of the quote I like is **"take it lightly."** After all we are just actors. **LINE PLEASE!**

"Life is a play that does not allow testing.
So, sing, cry, dance, laugh and live intensely,
before the curtain closes and
the piece ends with no applause."

Charles Chaplin

THE ACTOR (PART TWO)

If life is a play and the world is a stage, what role are you playing? Do you have a starring role, or are you an understudy? Do you stay in the background while others are in the spotlight? Is your part in the play smaller or less-than? If so, *do you want to be the star?* If you do, there are things you need to know. You have to be ready to take on the leading role. You will have to learn your part. You will have to give your all. You have to do more than dream; you have to do something about it. You have to be ready to get noticed; after all, if all the world is a stage it means people are watching. **"To be a star, you must shine your own light, follow your own path, and don't worry about the darkness, for that is when the stars shine brightest."**(Unknown)

If you are the star, you have to work more and work harder. *Still want to be the star?* If you are the star of the play of life that means you are out front most of the time, having to please the house. If you are the star you have to shoulder the responsibility for the play's success. They will remember you if the play was good, but they will be just as likely to curse you if it was bad. Remember it is the excellence that matters. *Still want to be the star?*

Truth is, being the star in the play of your life is not a choice; you were born into that role and will always play it. You may think your life is common or uninteresting; you may have many failures, or live in solitude or not have lived the life that others have lived. But your life is a magnificent and grand play being acted by the only actor that could ever play the part. I will never forget the first time I saw **Funny Girl**, I thought to myself that **Barbra Streisand** was born to play **Fannie Brice**. Well, you were born to play you, with your own successes and failures, your own dreams and desires. Your life (play) is unique, and you are one of a kind, and nobody can play you, but you. **YOU ARE THE STAR!**

"Dance like no one is watching.
Sing like no one is listening.
Love like you've never been hurt and
live like it's heaven on Earth."

Mark Twain

THE PLAY (PART ONE)

If all the world is a stage, and all the men and women merely players, where is the script? If we are here to act in the **play of life** how do we know what to do in each scene? So, where is the script? It is in our souls, and it was given to us at birth. However, the script is empty except for one word, **LIVE**; the pages are left blank for us to write on. The idea of a pre-written script for one's life would imply fate or destiny. Fate and destiny are excuses for things that would not be changed – things either loved or hated. We create our own fate and destiny by writing the script ourselves. **"It's choice, not chance, that determines your [our] destiny."**(Jean Nidetch) The pages are there for us to write our desires, our dreams, goals, wants, successes, and we unknowingly write our fears, failures and disbeliefs. But the script is up to us, and the play will not move along without it.

We write our script in many ways, much of it through the image we have of ourselves. That image can be interpreted as inner voices giving us our cues and directing our actions. We can allow those little voices from backstage (in our heads) to become our voice having us speak falsely and change the scene from joy to drama. The script is up to us because God gave us freewill. Freewill is that double-edged sword that cuts both ways. We can use our freedom to create bondage and play the part of the victim, or we can use it to be triumphant and play the part of the hero. It is not only done unto us as we believe, it is done unto us as we choose. We can misplace the blame and believe that our choices are being directed by outside forces, but choice still lies within us.

Although God is the Producer of the play, we are the playwright, the director, the actor, and the stage crew. We call for the curtain, clean the theater, turn out the lights and put the play to bed every night. The responsibility of the play lies deep within us, and few are lucky enough to know it. There are many scenes in great plays where the actor walks past a mirror and catches a glimpse of himself. In that moment, there is always a pause, and then everything changes; we instinctively always understand that the change begins first in the mind of the actor, in that moment the script changes. For us that mirror or reflection is everything that is being acted out in the play of our lives. If we do not like what we see we can call for a rewrite, but it is up to us. **PLAYED ANY GOOD PARTS LATELY?**

"We pass through this world but once. Few tragedies can be more extensive than the stunting of life, few injustices deeper than the denial of an opportunity to strive or even to hope, by a limit imposed from without, but falsely identified as lying within."

Stephen Jay Gould

THE PLAY (PART TWO)

The play is the vehicle for an actor to attain stardom, so if properly preformed the result of the play for the actor is greatness. Within the play, the actor has an opportunity to become everything he needs to be. The play allows him to be who he needs to be without sacrificing who he already is. **Cary Grant** said, **"I pretended to be somebody I wanted to be until finally I became that person. Or he became me."** He became Cary Grant, the movie star, without giving up the beliefs he had about life. In fact the beliefs he had were the path to becoming who he became, but he did not have to sacrifice his unique self to become Cary Grant. Staying true to himself allowed him to reach the heights of stardom and beyond.

If we reach for stardom and make it, it probably means we danced to our own music; sang our own song, gave all that we had, and loved a little along the way. It means that we did not settle for something less than our idea called for, or settle for another's ideas. We reached for something and touched it because the play is about possibilities. We did not shrink in moments of drama or give way to scenes of sadness or grief, but instead looked at the possibilities all around us. We were drawn to this play to act in it. We have inner strengths and truths that support everything we do. We all have moments of stage fright, but I do not think anyone has ever died from it.

The play is asking us to show LIFE; it is asking us to LIVE. It is asking us to stay true to ourselves and what we believe. It is asking us to look for the good and find it. It is asking us to try harder, when others are not. It is asking us to keep our word, even if others do not. Life asks us to believe in infinite possibilities in a finite world. **Alfred North Whitehead** said, **"Our minds are finite, and yet even in these circumstances of finitude we are surrounded by possibilities that are infinite, and the purpose of life is to grasp as much as we can out of that infinitude."** God does not hand us a script that says we have to be good, or worthy, faithful or kind, but if we write them in our script, we can experience them. If we believe that all things are possible we can seize any moment and choose to make it better. An unwritten script in the play of life is an opportunity to shape our experiences. It means that we can be ourselves and live life to the fullest without giving in to the status quo. **AND THE AWARD GOES TO!**

"I will not die an unlived life. I will not live in fear of falling or catching fire. I choose to inhabit my days, to allow my living to open me, to make me less afraid, more accessible, to loosen my heart until it becomes a wing, a torch, a promise. I choose to risk my significance; to live so that which comes to me as seed goes to the next as blossom and that which comes to me as blossom, goes on as fruit."

Dawna Markova

THE STAGE

In Genesis, we find, **"In the beginning God created the Heavens and the Earth."** If we read on, **"...thus the heavens and the earth were finished, and all the host of them. And on the seventh day God finished his work that he had done, and he rested on the seventh day."** God set the stage for the play of life by creating the heavens and the earth and the actors. A grand stage for the grand play of life where every actor has their own starring role, and within each actor lays the quandary from which we took our first cue. The play begins with the actor asking the question, **"Who am I?"** The human, the actor, the created, the child; who am I? With the stage set, the actor and the plot in place, the play begins.

The first stage set was lush and fruitful with everything that the actors could ever possibly want; where God (the producer) walked in the cool of the evening and watched over the play. But with one bite of an apple, the scene goes dark, and the set changes. With the set change, the plot becomes about suffering and hardship, lack, limitation and survival. No longer concerned with identity, but just making it through the scene, the actors change the plot to **"What is life."** But soon the actors found that the answer to what is life keeps changing, because life keeps changing. We live by universal Laws and put different things in and get different things out. So the actor, bewildered, realizes the original plot, "Who am I?," seems to be the right plot.

The **Greeks** gave us an idea of what the name of the play should be, **"Know Thyself."** Through their understanding of the *"gods,"* their tragedies, dramas, and great wisdom, they knew that self-knowledge was the key to the return. If we ever get back to the lush and fruitful beginning of human existence, it will be through self-knowledge. So the play goes on with a cast of millions, where the actors seek love, passion, greatness and joy; yet under seemingly limited circumstances.

The stage as we know it today can have pitfalls, trap doors, and dark places. But this stage is only a false front to the original stage that lies within. Within each and every one of us lies the original stage of life, the garden where our dreams and desires are realized. It may be a place that seems far away, but it is not. To find it, we have to find ourselves; to see it, we have to see ourselves. **Jesus** said, **"The kingdom of God is within you."(Luke 17:21)** The real stage, the original stage, is the kingdom of God and it is here and now, lush and fruitful. **CURTAIN!**

"Whatever It was that made this earth
the base, the world its life, the wind
its pillar, arranged the lotus and the
moon, and covered it all with folds of
sky with Itself inside. To that Mystery
indifferent differences, to It I pray..."

Devara Dasimayya

THE MASTER IS IN THE MIRROR

Have you seen him?
This master of all that is known.
I'm told the one who stands before him,
Would live life as if on a throne.

I was told I would find him here,
Here in this old broken down place.
And so I keep wandering these empty halls,
And yet, I don't find a trace.

Rambling from door to door,
In rooms bewildered, I walk alone.
Things seem to appear then disappear,
So nothing seems quite my own.

A chair, a pillow, a book to read,
Some time to think out loud.
A nomad inside a darkened place,
Away form the common crowd.

Sometimes I hear the sounds outside,
Or see movement beyond window's sill.
Shall I join them? No, I'll stay inside,
And seek the master still.

I am told I will know him when I see him,
Because he is the truth of me.
A sacred form that stirs my soul,
And my memory of eternity.

With wisdom in his pockets,
The past and future dance on his lips.
Beauty sleeping in his eyebrows,
Fate gripping his fingertips.

I have searched for him for such a longtime,
Yet my journey is still so cold.
What room, I wonder – where will he be?
When I find him, will I be too old?

What I do find is this dammed old mirror,
It's always blocking my view.
I'd look for the master behind it,
But it never lets me through.

Sometimes I just stand staring,
Wanting to see what could be.
A glimpse of what might be hiding behind,
But the mirror just stares back at me.

I think it's trying to show me something,
But what, I cannot know on my own.
So I keep looking for the master,
But yet, I still stand here alone.

"Funny thing about the moon,
How it keeps popping up, trying to see.
It goes round and around in circles,
Looking for God, just like you and me."

Chapter Eight:
"I Am That I Am"

THE NEW NAME

The spiritual experience for many can mean a new name. The Bible records many name changes: Abram became Abraham, Jacob became Israel, Paul became Saul, and Simon became Peter. All Popes take a new name; nuns, monks, and many devotees are given new names. Name changes in the Bible were usually given to establish a new identity; most of which were given by God after some heroic act or experience or to reward acts of faith.

Does this mean that because we are on a spiritual path we should receive a new name? No, I do not think so. But I do believe that if, through the spiritual path, we have found a new identity, that is, an identity that believes more, loves more, and feels more at peace with the world, then we have a new name. It may not be John, or Mary or Joseph, but a new identity by which the world knows us. Our new name is how we know ourselves and how we live in the world. Our new name is not something we sign on a check, but rather a new paradigm; a new way of knowing ourselves that reflects our relationship with God. We still are called by our common names, but nothing is common about us anymore.

"To all who long and strive to realize the Self,
Illumination comes to them in this very life.
This divine awareness never leaves them,
And they work unceasingly
for the good of all.
When the lamp of wisdom is lit within,
Their face shines, whether
life brings weal or woe.
Even in deep sleep they are aware of the Self,
For their mind is freed from all conditioning.
Inwardly they are pure like the cloudless sky,
But they act as if they too were like us all.
Free from self-will, with detached intellect,
They are aware of the Self even
with their hands at work.
Neither afraid of the world, nor
making the world afraid,
They are free from greed, anger, and fear."

From the Hindu text Yoga Vasishtha

THE DIVINE NAME

At the burning bush God announces to **Moses, "I Am That I Am."** There are many ways the statement has been translated: I Am That I Am; I Am Who I Am, I Am What I will be, and I Am the Living God. It seems to me that God not only announces that God is GOD, but at the same time announces the Great Law. The "I Am" is the Divine Name, and the statement I Am That I Am is a statement of being. And what seems to change is what comes after I Am. The I Am pronounces who we are, and what we claim after that is what we will be. The statement of being becomes the Law of our being. Every time we say, "I Am" we are invoking the Great Law, and we take it so lightly and use it so commonly because we do not understand the power of it, or the meaning behind it. **Thomas Troward** calls it the **Lost Word** because we have lost its meaning.

We use I Am to claim lack, anger, disease and every discord that comes into our lives, when we should be using it to claim the greatness of life: I Am Love, Joy, Peace, and Goodness. Make these statements today to yourself and reveal the power of the I Am. If we believe the Bible when it tells us that we are made in the image and likeness of God, then we must understand that that statement is a statement of our being as well. We are the individual within the universal.

"This 'Word' is always in our hearts, for the consciousness of our own individuality consists only in the recognition that I AM, and the assertion of our own being, as one of the necessities of ordinary speech is upon our lips continually. Thus the 'Word of Power' is close at hand to everyone, and it continues to be the 'Lost Word' only because of our ignorance of all that is enfolded in it."

Thomas Troward

THE MEASURELESS GAP

In Michelangelo's famous fresco **The Creation of Man**, also known as the **Creation of Adam,** it is clear that man and God are not quite connected. That seemingly small space between the finger of God and Adam's finger shows that if we try we might be able to touch God. It clearly points out that we have the ability; however, Adam's finger is pointing more at the earth and not at God. It seems to me that God is making more of the effort, and if we could but try a little bit harder and reach a little higher we could touch our maker.

Even though the distance in the fresco seems small, it is measureless, because it is different for each of us. That space or distance is the movement, or lift, or journey we take from our earthly human identity to our heavenly identity with God. It is the stretch we have to make from knowing ourselves as separate and alone to knowing ourselves as one with the cosmos. In truth, we are only a touch away from our divinity, yet that space keeps us in our humanness and our worldly experiences of lack and limitation.

"God moves in a mysterious way
His wonders to perform;
He plants his footsteps in the sea
And rides upon the storm."

William Cowper

THE MOON WAS ONCE A BUTTON

The moon was once a button
That held together God's shirt;
White silk, long sleeves,
Over a gauzy cotton skirt.

Back then we could look at God
As he went about his way;
Or reclining on his throne,
Just passing the time of day.

In those days it was good to see
What we worshiped from afar;
Like children watching a parent,
No matter how old we are.

Then one day something happened,
A curious thing back then;
God ripped off the shirt and skirt,
And we never saw God again.

He flung the shirt into the heavens,
And it became the Milky Way.
The skirt became the nebula,
The stuff of creation some say.

Oh, and yes, that button,
God tossed it in the air,
Flipped it like a quarter,
As if he didn't care.

As it flipped it started catching
The sun's golden rays of light,
Sometimes a tiny reflection,
Other times big and bright.

So the button became mysterious;
Its craters from threads ripped none too soon.
A nightly reminder of a time long ago
From a button we know as the moon.

Funny thing about the moon,
How it keeps popping up, trying to see.
It goes round and around in circles,
Looking for God, just like you and me.

"The waters will wash away from your heart
The memories you can't forget.
Pictures of all those yesterdays;
Broken dreams that you regret."

Chapter Nine:
Is Life A Gamble?

FAITH

Ever think that things just happen? If so, you might think that life is a gamble and that our lives depend on some cosmic **Roulette Wheel** for either fortune or failure. Sometimes life can seem like a gamble because we count on things, we know that something will come to us or we just know that something will happen, and it does not. Getting that new job, that raise, but something happens and things change, and so we feel as though we have lost, and wonder why. We feel as though we put everything on black, but it came up red; and life has handed us an unfair blow. It is times like this when we have to turn to our **faith.**

Ernest Holmes tells us that **"Faith is an inner knowing. It is real. It cannot die. We sometimes lose sight of it in stress, but it ever waits to be rediscovered and for us to put it to work."** Any moment we feel as though we have lost is the time to put faith to work. Faith causes the power to work for us and changes the loss to a win. Faith allows us to see that our desires are possible and moves us in ways that before we might not have thought possible. With faith, stepping into the unknown is not a gamble, but an assurance that the universe supports us.

Faith is a belief without logical proof; it gives us the confidence to move forward. Faith is trusting that the right answer to any problem will come to us. Within the experience of losing, faith becomes a powerful moving force. It tells us that what just happened is not the end. It tells us there is more; that if we just keep looking for the good we will find it. It tells us that the right people, places, and things will show up and will benefit us and our desire. Our faith becomes the Law in any experience, and the Law **MUST** produce.

Faith tells us that the only thing standing between us and our desire is consciousness and action. It is that something that tugs on our heart strings telling us that God works through each and every one of us. It is God in us saying, *"I CAN."* But God can only do for us, what God can do through us; we have to be willing to be moved; moved in mind, body, and soul. Faith allows us to be moved.

Faith is standing at the threshold of our desire and sensing God gently pushing at our backs. It is the universe whispering, **"Go ahead."** It is **"...taking the first step even when you [we] don't see the whole staircase."**(Martin Luther King, Jr.) Faith gives us the courage to go ahead and do something even though we still might be afraid. Faith does not mean we have annihilated fear, or denied it, but we are not immobilized by it. Faith is the action of taking the steps toward our goal. Faith takes the gamble out of life and allows us to open our lives to what God has for us. **HAVE FAITH!**

*"When you have come to the edge of all
light that you know and are about to drop
off into the darkness of the unknown,
faith is knowing one of two things will
happen: There will be something solid to
stand on or You will be taught to fly."*

Patrick Overton

LUCK & CHANCE

To say that life is a gamble would imply many things. Mainly, it would mean that we, or something (the universe, God), is spinning a wheel and whenever it stops someone would either win or lose. If you walk through a casino somebody is always winning, and somebody is always losing. Somebody always wins the lottery, eventually. Call it luck, fate, chance, but it is always happening. And, at one time or another, we all have won at something. We all have had moments when we felt as though **"Lady Luck"** was on our side; either with money or a new job, or something. The "win" could have seemed to have happened for many reasons: we just knew something was right; or we had a hunch; or we were willing to a take risk; or we were in the right place at the right time. And at that moment, life as a gamble might have seemed like a viable idea.

If we look at the idea as a possibility for life we have to ask ourselves, *"When we are born do we deposit two quarters in some cosmic slot machine and then pull the handle?"* If the answer is yes, we might find that we might have to wait all of our lives for the red, white, and blue 777 to show up and change everything; while, in the meantime we spend our time here on earth surviving on the occasional wild cherry. So, can we really rely on luck or chance?

The word chance can be defined as:

> The degree of probability that something will happen, or something will happen as the result of a combination of circumstances, or sequence of circumstances, or unpredictable and unknown factors coming together causing an event.

Sound complicated? Having our lives decided by **random number generators** would be equally as complicated. The possibility of our lives being determined by unknown factors taking risks with our emotions, love affairs, careers, money, health, happiness and our whole being is a very scary idea. **Richard Bach** tells us that **"Nothing**

happens by chance, my friend... No such thing as luck." Then why is it that some people seem to have all the luck?

The answer lies in consciousness. Everything that happens to us is a matter of consciousness, and consciousness is not chance or luck. If people feel they have the right to win at life, they will win. If they have self worth and feel deserving of something, they will get it. On the other hand, if one feels undeserving they will never win at anything. We can tell ourselves that we are lucky and deserving all day long, but where it will do us any good is at a level deeper than our everyday thoughts.

People who seem to win at life seem to feel good about themselves on this deeper level, without feeling that luck or chance had anything to do with it. They would probably be the first to say that life is not a gamble but a sure thing. They would also probably say that life is not always rosy, and that good and bad, happy and sad come to us all. We all live in a world of ups and downs, but life was never meant to hinge on the turn of a card or the roll of some cosmic dice.

"Depend on the rabbit's foot if you will, but remember it did not work for the rabbit."

R.E. Shay

SYNCHRONICITY

If you are on the fence about whether life is a gamble or not, consider those moments in life when everything comes together in a matter of seconds but have no explanation; experiences that cannot be explained by cause and effect. Are they chance or intelligence? Is it God, or do things just happen? Synchronistic experiences give us reasons to ponder these questions. **Carl Jung** coined the term synchronicity to describe what he called **"a causal connecting principle"** that links mind and matter and said that this underlying connectedness manifests itself through meaningful coincidences.

In an English class I was in some years ago, the class was given a short story and our assignment was to read it and be prepared for discussion the following class. The story had three main characters and a lot of dialogue. As I read the story I was compelled in the strongest possible way to highlight the conversations. I used a different highlighter color for each character so that I could instantly tell who was saying what at any given point of the story. After I was finished, I look at it and thought, what a mess! I cannot take this back to class and take the chance of another student or the teacher seeing it. **"Why did I do that?"** So I planned to get a new copy before the next class. I made two attempts and both times the teacher was not in. I left a voice mail and sent an email – and nothing. I wound up in class with the essay, having huge yellow, green and pink highlight places all over it. *What a mess!*

At the time I was about a year into my spiritual path and had found that I would have to have a college degree if I wanted to become a minister. I started the process at a local community college in Dallas. I was 52 and the rest of the students in this particular class were all under 30. I called it the class of 25 – 25 year olds, plus one. Talk about not fitting in. They were nice to me, but I stood out. The day of the short story, I stood out even more.

After our teacher had gone over some points about the dialogue in the story, he chose three people to read aloud the dialogue as if it were a real conversation, or they were in a play. He chose me to read the part of the main male character. My first reaction was, *"Oh no,"* I'm not going to be able keep up. However, as I looked down at the paper on my desk (here is the moment of synchronicity) I realized, **"I'm green!"** I do not have to figure out my parts. All I have to do is follow the colors down the page and speak-up when it is my turn. Because the speaking parts were not together the other two students fumbled to find their lines, misspoke, and even at times asked, **"Where are we?"** I was so ready that I even started responding before they would finish their parts, and at one point even received some applause when the tone between the characters got a little heated.

After it was over I got a lot of smiles, and the teacher made it a point to come up to me, and said, **"Good job,"** shook my hand, and told me that I was a good sport. On the other hand, the student sitting next to me said, *"Looking at your paper, you must have known he was going to do that."*

Truth is, no I did not know, but the universe (or God) did. Everything was brought together in a way that was meaningful to me. Because of the outcome, I was able to trust my inner feelings a little more. I was able to see that even though I tried to change things, and would have if I could have, it still worked out to my advantage. Everything worked together in a wonderful way.

Things like this happen, and we all have said, **"Well, you know what they say about great minds."** Well, there is a Great Mind, and It causes everything to happen. **"Synchronicity is choreographed by a great, pervasive intelligence that lies at the heart of nature, and is manifest in each of us through what we call the soul."**(Deepak **Chopra)** Synchronistic events reveal an underlying pattern of harmony and meaningful connections between the inner and outer world. Synchronistic experiences are key to our understanding that all things are connected and come from the One. It is God letting us experience, for a moment, the underlying order of the universe that connects time and experience in a brief moment of awe. We love for mysterious things to happen, and synchronicity for us is magic.

God, in Its Infinite desire for us to know, keeps putting together all the little bits and pieces of our lives and shapes the world around us, for us. Synchronicity is God's way of letting us know that nothing is chance and that everything in life is a sure thing and not a gamble.

"With one breath, with one flow,
You will know
Synchronicity

A sleep trance, a dream dance,
A shaped romance,
Synchronicity

A connecting principle,
Linked to the invisible,
Almost imperceptible,
Something inexpressible
Science insusceptible,
Logic so inflexible,
Causally connectable,
Yet nothing is invincible..."

The Police, 1983

FREEDOM

If we believe that **outside forces** impose limitation, hardship, suffering, and bondage on us, life can seem like a gamble; we never know what is going to happen. Another way of looking at it would be, believing in a vengeful God in the sky playing tricks on us. With the realization that God is Love, how could we ever think that God could be vengeful? But yet, there still seem to be those outside forces. How could those things keep happening to us if it were not because of those forces? **Ernest Holmes** says, **"We have thought that outside things controlled us, when all the time we have had that within, that could have changed everything and given us freedom from bondage."** We have been imposing bondage upon our experience ourselves by misusing Universal Law, but correct use of the Law will free us. Most of the time we are using this Law unconsciously, and unconscious use can be changed. It is not an outside force that keeps us in bondage to suffering and negative conditions, but ourselves. **Holmes** goes on to say that **"We are all bound, tied hand and foot, by our very freedom; our free will binds us; but as free will enables us to create the conditions which externally limit us, so it can uncreate or dissolve them."** Our limitation lies in the choices we make, the thoughts we think, both conscious and unconscious, and the actions we take. So does our freedom from limitation. Our freedom binds us, and our freedom frees us. But it is up to us.

God gave us volition and free will, and we are immersed in a Universal Law that creates by the very nature of our being everything that we are. Freedom is our birthright because it came from God; God would never impose bondage on us. God does not seek to control us or limit us. But we have to find out who and what we are and how these Laws create for us. We are never forced; God leaves us alone to discover for ourselves, and in the discovery we find that we cannot change the nature of these laws, but we can change ourselves. **"Man, then, is given power over his own life. He cannot alter the laws of nature, but he can so alter his relationship to them that that which had bound him may now free him."**(Holmes)

Disease, lack and limitation are mental images that are very real; however, they are not our truth. We are in bondage only to a false sense of the self; a self that says that life is limited, broken and cruel; a self that uses a great power unknowingly. At some point this consciousness was created; however, this consciousness can be changed. We can begin one by one to change the consciousness of things that seem like they came from outside forces.

"**Freedom means to eliminate from consciousness all those things which bind and limit the free flowing of the Divine Spirit through us and, at the same time, to exercise the faculty of personal choice.**"(Holmes) If we want to experience love, joy, happiness, or peace, we must choose these things in mind by thinking on them and becoming one with them.

There is nothing but freedom in the universe, and we use it every moment of our lives. There is nothing but Power in the universe, and we use it every moment of our lives. All life, health, peace, love, and all possibility exists because of this Power. But it is not the Power, but our use of the Power that creates any kind of bondage. It is our conscious use of this Power that will give us freedom. The highest truth is that we are SPIRIT, SOUL and BODY. We are made in the image of God; we are perfect, and we are perfectly free. The pathway to any freedom begins with the understanding that there is a Power in the universe that is greater than we are, and we can use it. **BE FREE!**

"My hope of freedom lies not in believing that there is a Reality and an illusion; but, rather, in staying close to Reality, and in thinking about those things I wish to experience rather than their opposite. Spirit, as Absolute Cause, and the material or physical universe, as effect, does not contradict each other."

Ernest Holmes

WHO'S PLAYING DICE?

Stephen Hawking once said, "All the evidence shows that God was actually quite a gambler, and the universe is a great casino, where dice are thrown, and roulette wheels spin on every occasion." People who live in the scientific world look at things very differently than the rest of us. I'm sure to a scientist, it could seem like God is playing **dice**. Life under a microscope shows chaos and uncertainty. But most of us do not look at life that way. We look at the big picture (not to say that science does not) where we do not see the randomness of life's behavior on the quantum-scale. But we do see and experience the randomness of life on the scale in which we live. Under the microscope of experience, we see the paradox of good and bad, happy and sad, love and hate, more or less, and one thing changing into another. So, does God play dice?

Einstein said, **"I shall never believe that God plays dice with the world."** He believed in a hidden variable theory (hidden order) and an underlying reality in all things. For me, the hidden variable is **belief**, and the underlying reality is **God**, and the words themselves give me hope. **Ralph Waldo Emerson** seems to disagree with Einstein about God not playing dice. In his essay **Compensation, Emerson** says, **"The dice of God are always loaded,"** and for him it seems the wins and losses are all about balance. He goes on to say, **"For everything you have missed, you have gained something else; and for everything you gain, you lose something else."** From this I recognize there is balance and fairness in the universe, or a level playing field where give and take comes without mistake or chance. He also said, **"The whole of what we know is a system of compensations. Every defect in one manner is made up in another. Every suffering is rewarded; every sacrifice is made up; every debt is paid."** This is God's hidden reality: A reality where beginnings and ends, time and space, seeds and fruits, causes and effects are the out-picturing of a Lawful Intelligence and Loving Creator.

Consider this. How could God have the Intelligence to create a universe where planets are suspended perfectly in place, where life evolves, and consciousness and space expands, and leave any of it to chance? If God does play with dice, the dice are always "loaded" and Universal Principles will win every time, and because we are made in the image and the likeness of... the odds are in our favor. We can and do win because the universe is rigged, if we will only learn the rules and play by them. As far as playing dice, I tend to agree with Einstein. I do not believe that God plays dice. I believe we roll the dice without realizing the consequences of our actions, and we win or lose according to what we believe.

"God does not play dice with the universe;
there is an order and reason for all of life."

Albert Einstein

CREATION'S SEA

There is a shore inside you,
Where the tides bring in what will be.
All of your tomorrows
Reach the edge of Creation's Sea.

This is a sea with endless shores;
A place that's never seen.
The edge of all your creations;
The realm of the in-between.

Here the light is still in the shadow.
The void contemplates the mass.
Spirit still choosing Its clothing;
The reflection not yet in the glass.

This is a place without time or space
Where silence is louder than sound.
The nothing between the something
Everything still unbound.

Here the winds will blow away
The things that are not you.
The things that were meant for others;
Those cards you never drew.

The waters will wash away from your heart
The memories you can't forget.
Pictures of all those yesterdays;
Broken dreams that you regret.

The tides will carry your courage;
The fear will be washed from your face.
Here your tomorrow awaits in twilight;
It's ready for you to embrace.

Call forth the winds of nature;
Be like the moon and swell the tide.
Creation's Sea is for you to command
From the shore that you have inside.

There is a shore inside you
Where the tides bring in what will be.
All of your tomorrows
Reach the edge of Creation's Sea.

"Riding the edges of space and time,
A cosmic tourist with an infinite pass,
Worlds upon worlds racing side by side,
As I Look through the Looking-Glass."

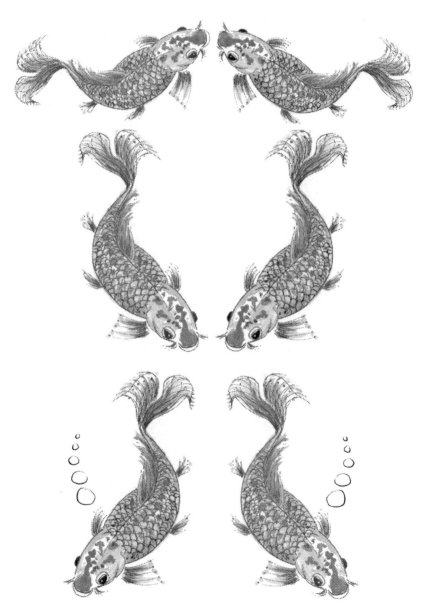

Chapter Ten:
Reality Check

A MONOTHEISTIC REALITY

*"Shema Yisrael Adonai Eloheinu
Adonai Echad." "Hear, O Israel:
the Lord our God is one God."*

Moses

If we look at humanity, the world, the universe, everything, and go all the way back to the beginning as far as we know it; if we look at all the sequences of Cause and Effect throughout time and history, we will arrive at a Universal First Cause or Source. One way we can do this is to look at the Bible. If we believe in God and believe (at least partly) in what the Bible says, we must believe that God is the originating Principle of and in all things, so there can be nothing else but God or Spirit. We will also find that this Source, God, or Spirit is **ONE**. There are many names for God in the Bible and elsewhere: Jehovah, Elohim, Allah, Bhagwan, Brahman, but only One God. Whatever name we give it, it is the **ONE Source**, the **ONE Self-existent** and **Self-transforming Power** in all Reality, **"There is none beside Me ... I am God, and there is none else" (Isaiah),** of which everything is some mode of manifestation: **"I Am That I Am."** If we continue this line of thinking we must come to the reality that there is but One Life and that life is our life too, and we are for a brief time (in this form) living a part of that Life. So we have to come to the conclusion that everything is God. **"God is the Seed; the Universe is the Tree, impulses and passions are the branches, intelligence is the flower, Pure Consciousness is the fruit, Love is the sweetness in the fruit."**(Sathya Sai Baba)

So, is God real, and what about us? Illusion or Reality? Or both? Are we living in a One Life Reality, or an illusion made up of stories from a Bible that we may or may not believe, or a dream that some god or gods have imposed on us? We are told (by some) that life is an **illusion**, if that is true does that make God an illusion? Even **Einstein** said that Reality is merely an illusion. However, if we go back to the conclusions that we have already made, we have to realize that God is very real and that makes us real, **"As Above, So Below."** God is not an illusion but a Reality. Life is not an illusion but a reality, not an empty dream to be filled with things that are not real.

I think on this plane of existence, right here where we are right now, life is as real as it gets, but there is an illusion. I think we make the illusion. The illusion is the masquerade or dream we live in – the illusion is that I am here and you are out there. **"We are one, after all, you and I..."**(De Chardin) The illusion is we make things appear to be what they are not. The reality is that this One Life is Perfect, Whole and Complete. God gave life Its All and All and over time we have clouded reality to fit our picture of life and not God's. In essence, we have taken the image we were made in and changed it; **that is the illusion**.

Illusions are but things that we experience but not the truth of us. The problem is that a lot of people make do with illusions and never really seek Reality. Some never fully realize the truth of their nature or wonder about the mystery of life. As for me, I do wonder and want to know. I want to understand Reality as best as I can. Wondering about the mystery of life can be an amazing path to walk. It is both external and internal, both the outer and the inner; a path where the two ultimately become one.

"Nations are many, but Earth is one; Beings
are many, but Breath is one; Stars are
many, but Sky is one; Oceans are many, but
Water is one; Religions are many, but God
is one; Jewels are many, but Gold is one;
Appearances are many, but Reality is One."

Sathya Sai Baba

A MIND REALITY

Do you remember the famous voice of **Captain James T. Kirk (William Shatner)** saying, **"Space... the Final Frontier"** at the beginning of every episode of **Star Trek?** Today we could say, **"MIND... the final frontier"** because many people are starting to explore the Mind from a personal standpoint. Today there is a growing need to know about the Mind and how it affects every part of our lives. Like space, we want to explore the Mind and everything about it. We want to go **"Where no man has gone before,"** in understanding; how it is that Mind is so vast yet so personal and powerful; so mysterious yet available; conscious yet unconscious; and universal yet individual?

Like the great minds in the exploration of space, many great minds have studied the Mind and from them we know many things; but there is still a great deal to know, if only on an individual level. How the Mind works can be universally understood (although mysterious), but how an individual uses their mind is another subject. We can use the Mind to create wonderful things, and we can be destructive at the same time. The last frontier may not be the discovery of the Mind Itself, but the discovery of ourselves and our use of the Creation Process.

First of all what is mind?

"What is the mind? No man living knows. We know a great deal about the mind, but not what it is. By mind we mean consciousness. We are now using it. We cannot locate mind in the body, for, while the body is a necessary vehicle for consciousness here, it is not consciousness. We cannot isolate mind. All we know about it is not what it is but what it does and the greatest philosopher who ever lived knows no more than this, except that he may tell us more of how it works." (Ernest Holmes)

If we are ready to wake up to the fact that we direct our lives through consciousness we can find book after book on Mind and Cause & Effect, positive thinking, and creation. We will find that it is all about being both human and Divine in a sea of Law. From Plato to Aristotle, from Jesus to the enlightened minds of today, we are told over and over again: WE CREATE OUR LIVES BY OUR

THOUGHTS. We live in this Mind and there is nothing but this Mind; we are individualized centers of this Mind; we eternally think into It, and It eternally creates. Because of the Mind's receptivity and Its Unconditional Power, the moment we think, something happens. The basic principle sounds simple – everything that is visible and everything that we experience is an effect of something that occurs in the invisible. The something we may never know, but we experience it all the time.

To complicate things, there are those two names, **Conscious** (or Objective) and **Unconscious** (or Subjective). Two minds? No. Not two minds, only two modes of operation. We know the stuff we think about on the conscious level, but we often do not know what is below (unconscious). This is where it becomes challenging. How we know what is below is a matter of taking a look at what is above – the visible manifestation of our lives. What do we like and what do we dislike? What do we want to keep and what do we want to change? If we are unconsciously drawing something into our lives that we do not like, it is coming from below; but we do our work from above, meaning our conscious mind. If we do not like what is happening in our lives, we use the conscious to change the unconscious. **Ernest Holmes** says, **"It has been proved that by thinking correctly and by conscious mental use of the Law of Mind, we can cause It to do definite things for us."** Our minds are tools and, if we could learn to use them, we could create better, happier lives. Through conscious use we can change negatives into positives, lack to abundance and sickness to health. The **final frontier** is learning we are powerful beyond belief, then using that power for good in our own lives – the life in the world, and beyond. **BEAM ME UP, SCOTTY.**

"Whatever the mind of man can conceive and believe, it can achieve. Thoughts are things! And powerful things at that, when mixed with definiteness of purpose, and burning desire, can be translated into riches."

Napoleon Hill

A MIRROR REALITY

Life as a Mind Reality shows us that our thoughts create the experiences in our lives, and that we and we alone are responsible for what is going on with our lives. Another way to look at this and help us understand it is to think of life as a mirror. **"Life is a mirror and will reflect back to the thinker what he thinks into it."**(Holmes) If we do not like what we are experiencing, we can change it, **"Change your thinking change your life."** The Mirror Reality is a Universal Principle that creates both good and bad in our lives. Our habitual thoughts are reflected back to us as the world we live in, the people we interact with, and the experiences we have. **Michael Beckwith** says, **"You attract to you the predominant thoughts that you're holding in your awareness, whether those thoughts are conscious or unconscious. That's the rub."** The Mirror Principle works in our lives as the **Law of Attraction** and the Law of Attraction says, *"What you are is what you will experience."* We are our thoughts, our beliefs and the universe does nor see us any differently. A physical mirror has no opinion and neither does the Law of Attraction. It is impersonal and only attracts like to like. The mirror does not make judgments about us as good or bad, nor can it say **NO**, you are really not like that, or you do not want this or that. What we hold inside makes the decisions and gives the directions.

Ernest Holmes says, **"When we stand in front of a mirror it reflects our image automatically, does it not? The reflection is identical and completely corresponds with the object in front of it. The reflection in the mirror did not put itself there."** We see things and experiences as coming to us from outside sources when in fact we are calling the shots; life is an inside-out job. Our thoughts (conscious or unconscious) give the Law the directions and become the script for the movie that our lives play out, but scripts can be changed. We can change any scene we are acting in, and we can even change our role in the movie. When we realize that life is the result of orderly laws and that these laws work for us and not against us, we can create happier lives by conscious use of them. God did not create a world for us to live in that is limited. There is a way, and we have complete access to it. When we realize that God and the universe are on our side, we can stand in front of the mirror and look at things a little differently and, therefore, experience life differently.

*"Reality is the mirror of your thoughts.
Choose well what you put in front of the
mirror. Change the mental movie that you
keep viewing in your mind to one that you
like. Keep playing it in your mind, and before
you know it the movie turns into reality..."*

Remez Sassion

A METAPHYSICAL REALITY

Metaphysics literally means **"what comes after physics."** This confusing word was initially used in the context of **Aristotle's** writings, **"First Philosophy."** Aristotle himself did not call these works Metaphysics. The prefix **meta** is commonly understood as "beyond," also "after," referring to Aristotle's work that followed after the chapters on "physics." However, the word beyond gives us a better picture of what we are trying to convey when we use the word metaphysics. It can be understood as **"beyond the physical."** Metaphysics investigates Universal Principles beyond the physical proof of science. Science works through logic; metaphysics does not. Logic appeals to the left side of the brain while metaphysics appeals to the right side and looks at the bigger picture and answers the bigger questions. Such as, what is existence?

If you are drawn to the metaphysical path you will get the opportunity to study the nature of God, causation, form and matter, and the relationship between mind and body. Anyone who studies metaphysics could be called a metaphysician. Opinions about metaphysics vary; **Kant** said, **"Metaphysics is a dark ocean without shores or lighthouse, strewn with many a philosophic wreck." Voltaire** adds, **"When he that speaks, and he to whom he speaks, neither of them understand what is meant, that is metaphysics."**

The study of life reveals to us that God is all and all. From metaphysics we learn that everything visible came from the invisible, and there is one Infinite Life closer than our hands and feet. We learn that Life and God are flowing energy within everything at all times. We learn it is within matter as well as space, and that there is no distance between this energy and our own. **"We're fields of energy in an infinite energy field."**(ee Cummings) If we want to get in touch with it, all we have to do is become aware of It. We already live in a metaphysical reality. Learning about it can only help answer our questions.

"Look out into the universe and contemplate the glory of God. Observe the stars, millions of them, twinkling in the night sky, all with a message of unity, part of the very nature of God."

Sathya Sai Baba

A Magical Reality

As a boy, I remember **Peter Pan** and **Tinker Bell** and a trail of twinkling pixie dust that made everything magical. In my wonderful make-believe world as a child I had some of that dust, and I would use it to make amazing things happen. I got everything I wanted and could even fly; at least in my childlike fantasy. As I grew older the dust naturally lost its power and slowly disappeared, but I never really lost the idea of making things happen magically. Some gesture or movement of the hand, the wave of a wand, uttering some ancient words; the dust was gone, but the belief was not.

I think we are all like that. I think we are all looking for some sort of magic in our lives to make the adventure happier and maybe even fly a little. The notion of being happy and getting what we want in life seems to be what life is all about. We are all on an adventure looking for those ends. We all want our needs met: good health, money, friends, love, and on, and on. We can find the promises of magic everywhere – there are many courses to take, books to read, people to tell us what to do; and we are told our dreams could come true if we would... But, nobody has the dust, or, at least, it seems that way.

When I found my spiritual path I was reacquainted with the idea of magic. Like other things, spirituality offers what would seem like magic to a lot of people. I was told (and I do believe) that I could have everything I wanted and that my dreams could come true, but it was not magic. It was work. The work was changing the beliefs about life that had changed since I was that child and had that dust. In essence, I had to become that child again, **"Truly I say to you, unless you change and become like children, you will not enter the kingdom of heaven."**(Matthew 18:3 - Lamsa) To make magic happen I had to go back to believing in magic. I had to let go of the limited ideas I had come to believe about life. In healing the false beliefs I found that the Kingdom is here and now, and not some far off kingdom that I had to fly away to, but I did learn how to fly. I learned that, **"All our dreams can come true - if we have the courage to pursue them."**(Walt Disney) I learned that all the magic I could ever hope for was in me and all around me. I learned that this kind of magic was not slight of hand or trickery or illusion, but the wonderful things that come with belief. I learned that good things were not out of the ordinary, but the way things should be. I learned with faith and God all things are possible, even magic. I also learned that life was meant to be bold and daring, and that each of us has the ability to succeed and make our own magic. German writer, biologist, theoretical physicist, **Johann Wolfgang von Goethe** said, **"Whatever you can do, or dream you can, begin it! Boldness has genius, magic, and power in it."** He also said, **"Magic is believing in yourself. If you can do that, you can make anything happen."**

So what about that dust? When I think about it these days I wonder – If God made the dust, and we are made from the dust, are we **that** dust? Are we, or can we be, the magic that makes things happen? I think the answer is yes. I think we are the magic. I think we create the magic because the magic of life is within us. I think we are here to say the words, make the gestures and sprinkle the dust; that is, the truths we know about God and life. I think we are here to not only learn how to fly, but to show others it is possible to fly. After all, **"All you need is faith and trust... and a little bit of pixie dust!"**(Peter Pan)

"Being born on the earth is the highest honor and greatest privilege. To be alive as human beings gives us the chance to pull off exquisite and Herculean feats of magic that are not possible in nirvana or heaven or any other so-called paradise, higher dimension, or better place."

Rob Brezsny

PARALLEL UNIVERSES

Am I waking up in the same world
As where I went to bed?
Have I dragged along behind me
Everything I've seen or said?

Perhaps I'm waking up here,
Because things changed back there.
Something happened, did I move on?
Right now I'm unaware.

Did I make that dash again
Across density's endless shores?
The ride I take, or does it take me,
Knocking on sacred doors.

Riding the edges of space and time,
A cosmic tourist with an infinite pass,
Worlds upon worlds racing side by side,
As I Look through the Looking-Glass.

Through tiny tubes of Reality
My journey in the blink of an eye.
Down a rabbit hole like Alice,
A black hole in my midnight sky.

The spin of all my memories
As I reach for the ring once more,
Birthing again this familiar body,
Appearing as it was before.

The persistence of my existence
Life's endless, deathless grasp,
The grip holding my identity,
Survival's forward pass.

This ride between my mingling worlds
Holds everything real for me.
The past, the future, this moment now,
Through the keyhole of eternity.

Am I waking up in the same world?
The question unanswered in me.
Even those who know the secrets,
Of this, what little can they see?

The answer is in the riddle,
What does the Circle say?
"Infinity lives inside Me,
I made Reality that way."

"Your eyes alive in butterflies,
Tulips are the strands of your hair.
Your smile in daffodils and marigolds,
Your footprints found everywhere."

Chapter Eleven:
Inspiration, Service, & Peace

INSPIRATION

The world's greatness and beauty are the result of **inspiration**. Inspiration affects every aspect of life as we know it: religion, politics, literature, art and economics. Inspiration moves people and nations to actions and purposes beyond the common and the ordinary.

Most definitions of inspiration include: a sudden idea or feeling, the physical act of breathing in or inhaling, the drawing of air into the lungs, breathed upon, to be in Spirit, a divine influence, or a sacred revelation. Inspiration also conveys the idea of motion, direction, or inclination into or to a place or a thing; the stimulation of the mind or emotions to a high level of activity; elevated energy and enthusiasm. I like the definition: **"in Spirit,"** having its cause from the Divine. So, for me, inspiration is the Spirit in the mind. It makes us think on a higher level. It gives us a breath of life and moves us in ways that we may not have moved before. We all have had moments we were **"Stung by the splendor of a sudden thought"(Robert Browning),** and inspiration gave us or the world something new.

Inspiration comes to us in many ways from many things: from a beautiful painting, a book, a poem, a play, a movie, a song, being in nature, at work or play, when we are still or alone. **Pablo Picasso** tells us that, **"The artist is a receptacle for emotions that come from all over the place: from the sky, from the earth, from a scrap of paper, from a passing shape, from a spider's web."** Even in our darkest hours, inspiration can be the one thing that changes everything. Inspiration is the key to the door that we did not know was locked, and once unlocked, we find newness, life, and genius, something grand. **"To the artist is sometimes granted a sudden, transient insight which serves in this matter for experience. A flash and where previously the brain held a dead fact, the soul grasps a living truth! At moments we are all artists."**(Arnold Bennett)

When the mind is stimulated great things happen; our God extended powers allow us to be the voice or the hands behind the birth of something wonderful. **"Poems greater than the Iliad, plays greater than Macbeth, stories more engaging than Don Quixote await their seeker and finder."**(John Masefield) We do not even have to be skilled in what our minds are telling us about, or know how it is going to happen. Inspiration, like the seed of an oak tree, contains everything that it needs, but we have to act. Inspiration can be like a dream or a shadow that can move very quickly, and so we have to catch it in the moment that it is with us and do something about it. **Henry David Thoreau** advises, **"Write while the heat is in you. The writer who postpones the recording of his thoughts uses an iron which has cooled to burn a hole with."** I have lost inspirations by not moving fast enough because my over cluttered mind had too many things to think about while I could have moved onto something better.

Inspiration is the difference between the common mind and the Divine Mind. It is the difference between limited and limitlessness. Our everyday minds are running all the time with idle chatter that is mostly non-productive. Some say we have something like 60,000 thoughts a day, but how many of them move us or change our lives? Inspiration breaks the chain of the mundane and allows the excellence of Spirit to pour in, but we have to be open to allow it to happen. **"Minds are like parachutes, they only function when open."**(Thomas Dewar) We have to let inspiration move and have its way with us. When we do, it will flow through us effortlessly, **"I did not have to think up so much as a comma or a semicolon; it was all given, straight from the celestial recording room ..."**(Henry Miller) However, if we do not have it, it is hard to find. **Freud** says that, **"It's** up to each of us to get and stay inspired. When inspiration doesn't come to me, I go halfway to meet it." **Ralph Waldo Emerson** adds, **"The torpid artist seeks Inspiration at any cost, by virtue or by vice, by friend or by fiend, by prayer or by wine,"** so we can not be lazy; we do have to work, but sometimes we have to wait.

If you are being moved by inspiration, it might be your path right now. If so, there are some things to think about. **The Path of Inspiration** is about having enough faith to act on it when it comes. It is about being dedicated enough to spend the time and energy to pursue the ideas and impulses that come from inspiration. It is trusting that your impulses are right, and knowing the difference between acting out of your own little intentions and out of inspiration. It is about the shift between mind and movement, **Ernest Holmes** tells us that a **"Man's mind should swing from inspiration to action, from contemplation to accomplishment."** The path is also about trust and saying yes and doing things you may not have done before. The path is knowing that God pours it in, but we pour it out, even if we do not know what we are doing. **"When you do not know what you are doing and what you are doing is the best, that is Inspiration."**(Robert Bresson) It can mean sleepless nights, **"A good idea will keep you awake during the morning, but a great idea will keep you awake during the night."**(Marilyn Vos Savant) It is also about being able to let it go when we do not think it is there; it is knowing that the **"Muse"** in us all may not speak for periods of time, but we are never really alone. We must know that this **High Genius** is with us all the time, waiting for that right moment to serve.

"You need not leave your room. Remain sitting at your table and listen. You need not even listen, simply wait, just learn to become quiet, and still, and solitary. The world will freely offer itself to you to be unmasked. It has no choice; it will roll in ecstasy at your feet."

Franz Kafka

SERVICE

If you are reaching out to others in one way or another you are on the **Path of Service**. I think we all walk the path of service at one time or another at some point in our lives: taking care of a child or a parent, reaching out to a friend or co-worker, or a complete stranger. There is something within us all that wants to do, when others cannot. I think this path, like the others, speaks first to our hearts and then our minds. The strings that tug on our hearts can be tied to many things, and opportunities to be of service come our way every day. One does not have to be on a path to show kindness in the world, but even occasional kindness makes us aware that serving our fellow beings makes our own lives more meaningful. **Marian Wright Edelman** tells us that **"Service is the rent we pay for being. It is the very purpose of life, and not something you do in your spare time."** Ask anyone in AA about paying the rent. Being in service to newcomers is what keeps the old-timers sober, and they will tell you that it is an ongoing job. However, for them, it is not really a job, but service done with understanding and compassion, with rewards for everyone concerned.

I have many opportunities to talk with people about their lives and "finding" themselves, and I always pass on the words of **Mahatma Gandhi, "The best way to find yourself is to lose yourself in the service of others."** As we give ourselves, our time, talent, our love, even things we do not know we have, something happens to us. I have never yet seen it to fail, that a person doing for others, always finds strength and direction for themselves. As a minister, I also let people know that **"The service of man is the only means by which you [we] can serve GOD."**(Sri Sathya Sai Baba)

Service, like everything else in existence is all about the **Law of Cause and Effect,** and doing good for others will always find its way back home to us. **James Allen** indicates that service is a way for us to relieve our suffering. He says, **"Forget yourself entirely in the sorrows of others, and in ministering to others, and divine happiness will emancipate you from all sorrow and suffering."** In other words, do something for others, and something will be done for you.

The path of service is about action. The actions we do in service for others rekindle the notion of oneness and that nobody is ever really alone. When we do even the smallest of things for others their faith is renewed, not only in God, but in humanity as well. The **Buddha** said, **"A generous heart, kind speech, and a life of service and compassion are the things which renew humanity."** Creating a better humanity is what service is all about. It is also about letting go of, or not having any judgment about, the "how or why" things got to be the way they are for those in need. On the other hand, service is all about joy and happiness, not just for those in need but for those who give as well. Service is never a one way street.

Like the paths of peace, compassion, and love, the path of service has been walked by many, and love seems to be the driving force behind service as a way of life. Perhaps **Mother Teresa** clarifies best for us the difference between love and service. She says: **"Love cannot remain by itself - it has no meaning. Love has to be put into action, and that action is service."** Service is taking the love that we feel and doing something about it. However, it is still about the love. She goes on to say, **"It is not how much we do, but how much love we put in the doing. It is not how much we give, but how much love we put in the giving."** Her life is a living example of how we **"love our neighbor."** Through God's Love we are drawn to be our brother's keeper. If we, of ourselves, come to know that we are love then we must come to understand that we are never really **"out of service."** Love never ends, so service never ends. Any action of service is a call from love, and love will always call on a corresponding action, **"Every soul attracts its own."(James Allen)** The path of service does not have to be an organized path with daily routines and actions that require a time clock, but the simple understanding that when we look into the eyes of those in need, we are seeing our own, and as much as we do for them, we are doing for the God in us all.

*"Lose yourself in the welfare of others;
forget yourself in all that you do-this
is the secret of abounding happiness.
Ever be on the watch to guard against
selfishness and learn faithfully the divine
lessons of inward sacrifice; so shall you
climb the highest heights of happiness,
and shall remain in the never-clouded
sunshine of universal joy, clothed in
the shining garment of immortality."*

James Allen

PEACE

The Path of Peace begins with the realization that God is all there is, and in God there is no anger, no confusion. God is never disturbed, hurt, or aware of adversity, or anything less than Itself. God is peace, and as much as we turn to the Eternal Presence that is within us, we turn to peace. Within the Universal Mind there is order and harmony, and that Mind is our lives, although It waits for us to express It in our lives. **Ernest Holmes** says that **"Peace stands at the door of your consciousness and awaits your acceptance of It. However, It does not stand outside your door, waiting for entrance, so much as It stands inside waiting to be expressed in everything you do."** If we wish to express peace, we must direct our actions and intentions to come from a place of peace. We must think peace, we must be peace.

The path of peace calls for the deep realization that God is in everyone, seeing past the anger and hate in the world, to the underlining harmony in creation. It calls for a Spiritual understanding of the Law of Creation and how our minds create our experiences. It calls for the individual mind to be quiet and calm. It requires us to step out of the littleness of our ego and in to the bigness of God. Peace requires us first to make peace with ourselves. **"The mind that is always confused and distraught is not at peace; the mind that is continuously upset and agitated by the little, petty things of life is not at peace; it is at war with itself. It is only when the individual mind ceases combating itself that it will stop combating others."**(Holmes)

Peace like love was given to us at birth. It was not born from human struggle. It is a part of who we are, and that is the reason why we want it so much. In times of disharmony we feel as though something has been taken away. When we see others without peace, we ache because we know that there is another way.

If you are called to the path of peace, you will see that there are many winding roads, blind alleys and u-turns on your way. You will find that there are many who believe that peace is not attainable or even available in this world of so many differences. You will see that this path is well worn from the shoes of those who have gone before you and those who walk beside you, all seeing the same reasons with the same passion and desire. Along your way you will have to stop many times to pray, meditate, and take time to search your soul for what is real and what is not. You will have many reasons to stop and look in the mirror and ask **"Is it all worth it?"** And the answer will always be the same, **yes**. You will find that **"Nothing can bring you peace but yourself."** (Emerson) You will also find that, **"There is no way to peace. Peace is the way."** (Gandhi)

"Be still, still, in every way, and accept that Peace of God, knowing It now heals everything within you that hurts, and that Its calming action enfolds all your experience. Think Peace, feel Peace. Know that you are Peace, because you are a definite, specific expression of God manifesting as you."

Ernest Holmes

THIS SLEEPLESS DAY

This sleepless day has taken me,
And pulled me from out of my bed.
It called me out of that merry-go-round,
The dream that lives in my head.

It took me down your sacred path,
And through your garden door.
A key was with me all along,
I just never found it before.

Here the beauty is overwhelming,
Life bathed in sun and shower.
Your joy unfolds with every leaf,
Your fingerprints on every flower.

Your eyes alive in butterflies,
Tulips are the strands of your hair.
Your smile in daffodils and marigolds,
Your footprints found everywhere.

The rose is full of your fragrance,
Your teardrops nurture the vine.
The stem is drunk with your nectar,
The air is your breath sublime.

Your face in a million poppies,
They turn when I walk their way.
They tell me you know I'm here with you,
On this beautiful sleepless day.

Your God Garden all around me,
Awake in this juncture of time.
Standing still in the dance of eternity,
The sway of our hearts entwined.

This sleepless day in your garden,
For me what a golden delight.
Free from my dreaming dream,
My will and my soul take flight.

The wondrous wonder of the moment,
I never want a second to end.
But it seems the dream is calling,
My eyes grow sleepy again.

Once again asleep on the merry-go-round,
I'll awake as I was before.
Again and again still dreaming that dream,
Searching for your garden door.

"Can you bend around the circles?
Are you here to unlock the lock?
Can you laugh at destiny's warning
By turning back
the hands of the clock?"

Chapter Twelve:
Meditation & Wisdom

MEDITATION

We can be called to the **Path of Meditation** because of the benefits. The practice of stepping out of our everyday minds and into the vast void that we all are a part of, can make us **healthy, wealthy** and **wise.** When practiced sincerely, meditation can heal our bodies, calm our nerves, guide our lives and unveil the mysteries of life. Meditation is mysterious, and a lot of people talk about it mysteriously. Phrases like, **"This dead of midnight is the noon of thought, and Wisdom mounts her zenith with the stars,"** (**Anna Barbauld**) come to mind. For me, the author is saying, when the mind is dead of everyday thoughts, our ability to comprehend real knowledge is at its highest, and Wisdom is reflected to the mind of the beholder. Meditation itself is not mysterious, only its inner workings. **Jeremy Taylor** said, **"Meditation is the tongue of the soul and the language of our spirit."** Sitting in silence is one thing, but translating the language of Spirit is another. Meditation is an experience in and of the Unknown; the transition between meditation and outcome is where the mystery lies. But the transition happens.

One thing that makes meditation so mysterious is all the different ways of doing it and the different ideas about it. One person may be all about the **silence,** another the **breathing,** or the **position,** the **mantra,** the **mudras,** the **beads,** or the **focus,** I could go on. When we could just sit and be quiet, observe, reflect, and if we can, let our thoughts go. All of the various ways of meditation are just to keep us busy while God works, but we have to supply the time and attention for God to work. **Ernest Holmes** says, **"Spiritual experience comes in the stillness of the soul, when the outer voice is quiet. It is a quickening of the inner man to an eternal Reality."**

A lot of the mystery connected with meditation seems to be about the silence. In our noisy world, with our noisy lives, and our chattering minds and mouths, silence seems to be the one thing we can not grasp. Adding to that, there is a mystery in silence that few understand, and that is the experience of ONENESS. **Thomas Merton** says, **"When we have really met and known the world in silence, words do not separate us from the world nor from other men, nor from God ...."** The more of **"US"** that experience Oneness, the more **"I"** there will be. Silencing the noise in our minds, our bodies, our lives and our world, gets us in touch with the Silence and the Oneness of God.

How do we meditate in silence? Just do not talk, do not use words. That is OK and doable, but how do we silence our minds? How do we stop the never ending, ever turning cycle of activity in our heads? This is where all the other things come in. We CAN silence our minds through concentration on our breathing, mantras, or counting. It is like trying to go to sleep, just count the **sheep** and your **mind** will go to sleep; do something besides think about your thinking. If a thought comes in, honor it and let it go without wondering what it means or why you are thinking it. **"Empty yourself of everything. Let the mind rest at peace. The ten thousand things rise and fall while the Self watches their return."(Lao Tzu)** We live in an ocean of mind and our conscious minds are like the ceaseless waves and the churning tides, but beneath this endless commotion there is a calm that is unaffected by anything on the surface. That calm is meditation.

Setting silence aside, there are meditations done with thoughts and meditations done with purpose and intent that are done with actions. There is Yoga, Tai Chi, Sufi Dancing, Labyrinth Walking, Art, and Chess, golf, cooking, sewing, and, yes, even fishing. Life is filled with opportunities for centeredness, attention and awareness. Any action done with a mind centered in awareness is meditation. The kind of activity does not matter, but what matters is how much consciousness we put into it. **"Drink your tea slowly and reverently, as if it is the axis on which the world earth revolves – slowly, evenly, without rushing toward the future; Live the actual moment. Only this moment is life."**(Thich Nhat Hanh) I am reminded of the sacred movements of the hands of the **Geisha** in the art of the Tea Ceremony (Sado), every movement a purposeful gesture, meditation in action.

Meditation is an art, and the Path of Meditation is living an art filled life. **"Meditation is one of the greatest arts in the world. In other art forms we create a painting, pottery or music, etc. But in meditation we create a fully conscious human being. Meditation is one of the most creative and fulfilling experience."**(Jiddu **Krishnamurti)** Like an artist has control of the brush through conscious awareness, we have control of our minds through conscious awareness. **"Meditation means awareness, alertness, watchfulness, witnessing. Witness your actions, witness your thoughts ..."** **(Osho)**

The Path of Meditation is about going beyond the common mind and becoming the witness and the awareness. The path is about letting God plant the seeds that we are not aware of until harvest. The path is about realizing the body's need for material food and the soul's need for spiritual food. It is about using the mind as an instrument and being its master and not its slave. The Path of Meditation is the gateway to spiritual mastery, freedom, and enlightenment; not about the pursuit of some invisible imagined bliss. The Path is about seeing, watching, listening, non-attachment and being.

"The very essence of meditation is to be so silent that there is no stirring of thoughts in you, that words don't come between you and reality, that the whole net of words falls down, that you are left alone. This aloneness, this purity, this unclouded sky of your being is meditation. And meditation is the golden key to all the mysteries of life."

OSHO

WISDOM

If you are called to the **Path of Wisdom,** you will know it. It will seek you everywhere you are and in everything you do. Like the moth near the flame of a candle, wisdom is something that will not give-up or go away: **"Wisdom calls aloud in the streets, she raises her voice in the public squares; she calls out at the street corners ..."(Proverbs)** You will find as **Socrates** did that **"Wisdom begins in wonder."** You will wonder about the secrets of life, the world and how things got to be the way they are. You will find that the path of wisdom, like all other paths, leads to God. This very personal path allows the seeker to contemplate not only what is already known in both the ancient and the new world, in both the written word and the spoken word, but through the seeker's own mind. Ultimately the path leads the seeker to what is real. Real wisdom comes from the One Mind. In the **Science of Mind, Ernest Holmes** says that the Divine Mind contains all knowledge and wisdom but must have an outlet. To be an outlet, one has to first become an inlet; this is the path of wisdom.

The path of wisdom can draw us to the paths of other wisdom seekers and teachers. We can be drawn to the wisdom of **Jesus, Buddha, Lao-Tzu, Plotinus, Socrates,** and **Kabir.** Wisdom is all around us in everything: in the movement of water, the currents of the winds and the waves, in the stillness of evening, but we have to be in touch and aware to sense it. Through awareness we become one with the wisdom all around us. Like the moth to the fire we are drawn, but not consumed. We enter the fire, but our sense of the self is not destroyed; instead it is expanded and inspirited by the true Self. Awareness is how we become an inlet; *WHAT WE BECOME*, becomes the outlet. Wisdom changes how we live in the world. Wisdom is more than the knowledge of who we are; it is more about how we live who we are. **"Never mistake knowledge for wisdom. One helps you make a living; the other helps you make a life."(Sandra Carey)** Wisdom is how we apply knowledge. Knowledge fills our lives, but wisdom builds our lives. Further, the path of wisdom is being true to our inner-self, bringing harmony and peace into our lives and the lives of others. Wisdom is honoring humanity and caring about the rights of others. Wisdom is reflected in the way we treat others. It knows there is a Spiritual solution for every problem, and that if we allow it, our Spirit will guide us to that solution. It is the path that knows that with God all things are **"possible."** Wisdom is having courage. **"Without courage, wisdom bears no fruit."**(Baltasar Gracian)

Wisdom is the **precious liquid** that **Buddha** spoke of that will cleanse the horrors of the world. It is knowing that change begins with us, **"Changing the world begins with the very personal process of changing yourself, the only place you can begin is where you are, and the only time you can begin is always now."**(Gary Zukav)

Wisdom is not something we can just go out and get, but we find within ourselves, and we can peruse it with a conscious mind. We can become teachable; all of us have something to learn. We can learn to slow down and become aware of the invisible behind the visible. We can consciously draw upon the reservoir of the Infinite Mind through becoming one with IT. We can learn that **"Wisdom is not a product of schooling but of the lifelong attempt to acquire it."**(Albert Einstein) Wisdom is the never ending path because we can never know all of God on this finite plane. In the realization of wisdom's importance, we come to understand that it must be a part of all other paths. It was wisdom that laid the foundations of the earth, so wisdom is the foundation of our paths.

*"Blessed is the man who finds wisdom,
the man who gains understanding, for she
is more profitable than silver and yields
better returns than gold. She is more
precious than rubies; nothing you desire
can compare with her. Long life is in her
right hand; in her left hand are riches and
honor. Her ways are pleasant ways, and
all her paths are peace. She is a tree of
life to those who embrace her; those who
lay hold of her will be blessed. By wisdom
the Lord laid the earth's foundations,
by understanding he set the heavens in
place; by his knowledge the deeps were
divided, and the clouds let drop the dew."*

Proverbs 3: 13-20

CAN YOU REACH FOR THE END OF ETERNITY?

Can you reach for the end of eternity
And embrace what comes to pass?
Can you stop the sand from falling
Inside the hour glass?

Can you encompass every act of joy
As well as all the sorrow,
Stand inside both good and bad
And face another tomorrow?

Can you hold the end of this moment
And claim it in everyway?
Can you love the night sky's beauty,
And still praise the sun each day?

Can you bend around the circles?
Are you here to unlock the lock?
Can you laugh at destiny's warning
By turning back the hands of the clock?

Can you reach beyond the darkness,
With you will the fight begin?
Will you hold the angel till daybreak?
Will you stay; choose this day, lose or win?

"The dawn has kissed the twilight,
Sweetened lips have touched once more.
Birthing this new horizon,
More glorious than ever before."

Chapter Thirteen:
Oneness & Joy

ONENESS

Although we seem to live in a disconnected world where everything is separated, everything is really connected, but not in a way that is visible to the naked eye. There is another eye that sees this interconnectedness. That eye is not the physical eye, but the consciousness of the Eternal "**I**," and it sees by the Light of Reality. The idea of Oneness is found in all religions, traditions, spiritual teachings and wisdom books, "**What is here is also there; what is there, also here.**" (Katha Upanishad) "**In the ocean of being there is only one. There was and there will be only one ...**"(Ashtavakra Gita) "**I and the Father are One.**"(Bible) Oneness wisdom seems to have a way of pointing at something out there to show that it is really in here. There is One Indwelling Presence in everything, and anyone on the path of this awareness must live in the paradox of looking at **otherness,** and seeing **sameness;** this path is the practice of rising above the illusion of separation. **Albert Einstein** said that we experience life as separate, and our task is to free ourselves from this "**self-imposed prison, and through compassion, to find the reality of Oneness.**"

The concept of Oneness conjures up the idea of advanced enlightenment or something understood only by Saints, Sages, Mystics and wise beings who do not seem to live in the world with you and me. But the very thing they teach says that it is all of us, not just some of us, and that we could see it if we looked for it. We are all created out of and live in the same *stuff,* so we are all in this experience together. **"We all drink from one water, we all breathe from one air, we rise from one ocean, and we live under one sky."(Anwar Fazal)** The stuff is Source, or Primary Cause, or energy, or God. **"It may be given a thousand names such as The Primary Cause / God / Energy / I. All that is created has its Self this Oneness."(Sri Sathya Sai Baba)**

We live in an ocean of Oneness, and for a brief moment, we have stepped out of the ocean and see ourselves as something other than it, separate. Oneness is a matter of getting back into the water and overcoming the idea of separateness. This is what the mystics of the world have done and invite us to do. When we are out of the ocean, or Oneness, there is something within us that wants to return. It wants to return, because that is where it belongs. We feel this same way when we longingly look at the stars. We feel what we do because we are made of the same stuff, and, at that moment, Oneness wants the separation to end. Getting in touch with this feeling constitutes the mystical experience.

The key to this mystical achievement is **"identity."** We have to know ourselves to know the ocean. **Know thyself** was inscribed in golden letters at the lintel of the entrance to the **Temple of Apollo at Delphi**, and the same thing is invisibly inscribed in golden letters over the entrance to all mystical paths. Oneness is the ability to rise above the idea of individuality (without losing one's self) where, **"We find ourselves, not absorbed, but immersed, in a Universality, each one being a unique, individual and different manifestation of that which itself is one, undivided, indivisible, and whole."(Holmes)** In oneness we rise above the idea of a limited self, **"Arise, transcend thyself. Thou art man & the whole nature of man is to become more than himself."(Sri Aurobindo)** The *MORE* is the realization that others are the same as one's self, and what lies in one heart, lies in all, **"In the fullness of one's spiritual realization one will find that He [God] who resides in one's heart, resides in the hearts of others as well – the oppressed, the persecuted, the untouchable, and the outcast."(Sri Sarada Devi)**

Those who dwell on the Spirit on the inside must also dwell on the Spirit on the outside, making the outer-most and the inner-most One.

156

The Path of Oneness can be walked in a cave, but it is what is outside of the cave where the real work lies. The path is the taking of all that we see, everything out there, all that we love or hate, everything that we agree or disagree with, all the forms of good and bad and taking it inside. We have to remember that God makes the sun to rise for us all and sends down the rain on the just and the unjust, because all are alike, **"We are one, after all, you and I. Together we suffer, together exist, and forever will recreate each other."**(Pierre Teilhard De Chardin)

Seeing the sameness in ourselves and others is all about **SURRENDER,** surrender to an acceptance of not only who we are, but an acceptance of who *"they"* are. Like many other paths, oneness is about love. It is the realization that even though someone may be doing wrong, inside that person is the same soul stuff, **"Try to love someone who you want to hate, because they are just like you, somewhere inside, in a way you may never expect, in a way that resounds so deeply within you that you cannot believe it."**(Margaret Cho) This kind of love comes from a complete surrender to the "I" that we truly are. The **I** that is able to see all life as one and God as One. The Path of Oneness is seeing the differences in the world without blame or judgment. It is about seeing differences as different modes of the One Being, **"All differences in this world are of degree, and not of kind, because oneness is the secret of everything."**(Swami Vivekananda) The path is about sharing what we know with the disconnected and widening the circle of truth.

"What I mean by the Principle of Oneness is this: That we must learn to realize that there's nothing separate or apart. That everything is part of everything else. That there's nothing above us, or below us, or around us. All is inherent within us. Like Jesus said, 'The Kingdom is Within.'"

Eden Ahbez

JOY

We can be drawn to a path in many ways. We can be drawn by attraction, curiosity, or simply wanting to connect with others. Something can look fun or interesting, like joining a Sufi dancing group, or learning how to pray or meditate with beads. Many paths are about the things we do not think we have, and sometimes we can want those things badly. Funny thing about that last part, most people who want something so badly, after they get it, they wind up giving it away or helping others get it. Ask any alcoholic about the **Path of Sobriety**, and they will tell you after getting their sobriety, they spend the rest of their lives helping others to get what they have. Joy is very much like that, because we can want it so badly. We can see that other people have it, and like the alcoholic and sobriety, think that it is not possible for us, but it is. In fact we should have it. Joy, like sobriety, is a natural state attainable by all.

In his poem, **The Same Stream of Life, Bengali Poet Rabindranath Tagore** tells us that God shoots in joy through the dust of the earth. That dust is not just the ground but also you and me and all that we see and do. That dust is our lives, and joy is a quality of our Spirit. Joy can be thought of as cheerfulness, but it is much more. It is more than happiness too. Joy is God Itself breaking through the crust of our lives into each and every experience. But joy must be recognized and embodied before we can experience it. Joy is, as life is, a blend of ups and downs and laughter and tears, but even in the midst of sadness or strife this inner sense of God prevails. That is why **Jesus** told us to **"be of good cheer,"** he knew that behind life's stage, behind all the drama, there is joy.

"And Joy is Everywhere;
It is in the Earth's green covering of grass;
In the blue serenity of the Sky;
In the reckless exuberance of Spring;
In the severe abstinence of gray Winter;
In the Living flesh that animates
our bodily frame;
In the perfect poise of the Human
figure, noble and upright;
In Living;
In the exercise of all our powers;
In the acquisition of Knowledge;
in fighting evils...
Joy is there Everywhere."

Rabindranath Tagore

THIS DAY IS LIKE A RUBY

The dawn has kissed the twilight,
Sweetened lips have touched once more.
Birthing this new horizon,
More glorious than ever before.

This day is like a ruby set in a fine ring of gold,
Encircled by heaven's delight.
Resting on the hand of the Holy,
Lustrous, wondrous, and bright.

This day awakens every treetop,
Giving life to flower and weed.
Calling forth the birth of creation,
The breath in the mouth of the seed.

This day is like a ruby set in a fine ring of gold,
A gem in a halo of fire.
Like the moon embracing the sun,
Eclipsing its heart's desire.

The brush has painted the mountains,
The vapors drawn from the sea.
The winds are telling their stories,
They're calling for you and for me.

This day is like a ruby set in a fine ring of gold,
What care of struggle or money?
Go out my friend and find you some bees,
And taste for yourself some honey.

"You are a Shining City
Your arms forever unfurled.
Your dream cannot be hidden;
Your light is the light of the world."

Chapter Fourteen:
Holiness, Passion & Spirit

HOLINESS

Holiness is a very *BIG* word. It brings to mind people that are considered Holy, the **"Holy Ones"** that is: Mystics, Saints, Sages, Monks, Priests, and Popes. We look at their lives as very different and very much set apart from the average person. As we know, the word **Holiness** can even be used as a reverent title that sets one apart from all others. And it seems to follow that if the person's life is holy their actions would be holy as well; therefore, their actions would also be set apart. We not only see people as being holy but things and places as well; we see shrines, statues, paintings, images, temples, cathedrals, and books: the Bible, the Koran, and Bhagavad-Gita as being holy and set them apart, too. In all, holiness seems to set someone or something apart. **Kadesh** is the Hebrew word for Holy or holiness, and it basically means something that is set apart or separate or something that is elevated out of the ordinary; something *"special."* The Holiness of God is certainly something set apart or seems to be set apart from the world. We not only struggle with the image and idea of God but things holy as well, because they have been set so far apart from life as we know it. It seems only those extra ordinary lives manifest or experience what we call Holy, yet holiness is something that we all possess.

Meister Eckhart the wise German mystic said that **"We should not think that holiness is based on what we do but rather on what we are, for it is not our works which sanctify us but we who**

sanctify our works." I believe he is inviting us to understand that actions or things of themselves are only as holy as we make them. It is who we are and where we come from that creates holiness. He also says, **"Whoever possesses God in their being, has him in a divine manner, and he shines out to them in all things; for them all things taste of God and in all things it is God's image that they see."** It is not the object we see that is holy, but the eyes we see it with that makes things holy. Furthermore, we would not even be able to comprehend others as being holy if we did not have something holy within ourselves. This is the idea behind the concept which says, **"What we are looking for, we are looking with, and we are looking at."** Possessing God in our being makes us holy. It is because of whom we are that we are able to choose our actions, and actions that come from holiness cannot create harm, only goodness.

The **Path of Holiness** has to be based in the notion that God is the essence of our being regardless of what we call God; regardless of race or religion or tradition it still works the same. The God in us may be called many things: the **Christ** nature or consciousness, or **Buddha** or **Krishna** nature, or **Higher** consciousness, but whatever we call **IT**, **IT** wants to live Its life, and Its life is Holy. **"Every creature is a word of God,"** **Eckhart** says, and that word is Holy. In essence, for me, the Path of Holiness is all about the individual life ending the idea of setting people and things apart and bringing everything together. Living a holy life should not put us or our actions on a pedestal, but rather level us and our actions to the earthly plane of the here and now where the work is to be done. It is all about being in the world **"but not of it."** The only thing we need to set apart is our faith and consciousness, keeping them above the conditions we see and experience in the world.

The path of holiness is about bringing the outer and the inner, the higher and the lower, together in one's life enough to understand that life is about more than one's own, and that other lives depend on what we do. Even in a cave, holiness must serve beyond one's own needs and comforts. Being holy has always been about others. **"You are not here in the world for yourself. You have been sent here for others. The world is waiting for you!"** (Catherine Booth) If we look at the lives

164

of those we consider holy, such as Jesus, we will see in every instance that their lives were about the bigger picture; even after their lives were over, their holiness still serves. This is the way of holiness.

Every path requires one to be empty so that one may be filled, and the Path of Holiness is no different. It requires an emptying of the less than ideas of the outer life, so that one might be filled with the image of God. This path requires a deepening of one's inner life. For this path, living in the world comes from heart and depth. It requires one's commitment to the practice of communion with God in prayer and meditation. It is being able to stand in the means while others are only reaching for the ends. The Path of Holiness is simply living life as if it belongs to God; it is holding the vision of God in all and sharing the Divine Blessings of Truth and Love. One does not have to be a Messiah, or a Pope, or a Monk to walk this path, but have a love that reaches beyond the bounds of everyday life.

"As God delights in his [Its] own beauty,
he [It] must necessarily delight in the
creature's holiness which is a conformity
to and participation of it, as truly as [the]
brightness of a jewel, held in the sun's beams,
is a participation or derivation of the sun."

Jonathan Edwards

PASSION

If you are ready to take a particular path, do you feel on *fire* about it? Do you have a burning desire in your soul to do it, or have it, or be it? If you feel the fire, you can thank the **Titan, Prometheus**. In Greek mythology Prometheus was the god who stole fire from Zeus and gave it to mortals. We all have that fire in us, and although it has been called other things, for me, no word describes it better than **PASSION**. It is the driving force or the burning desire behind the things we are called to do or have. As the myth goes, Prometheus paid a very high price for giving us the power that propels us to do great things and be great people. Passion is the fire of life that moves us in ways that only a myth could even begin to explain. There are many, many passions in the world. Music, art, acting, dancing, writing, the passion of giving, the passion of belief, the passion of courage; we can be passionate about almost anything. And we do not even have to have a particular talent or know what we are doing, **"All you need is passion. If you have a passion for something, you'll create the talent."(Yanni)** All we need is that fire.

Someone's burning desire was the beginning of all great creations. Making something is one thing, but making something with fire is another. Just as knowing how to do something is one thing, but doing it with fire is another. **"A strong passion for any object will ensure success, for the desire of the end will point out the means."** **(William Hazlitt)** Passion is the difference between something bland and something that has sparkle. The fire on the inside makes it come to life on the outside. Passion is the difference between sleepwalking and living a dream. Passion makes us come alive, because it allows us to do the things we are called to do. **Harold Whitman** said, **"Don't ask yourself what the world needs; ask yourself what makes you come alive. And then go and do that ..."** Passion not only gives us life, but it makes our lives meaningful and fulfilled as well.

One could ask, **"Is passion a path?"** I would say that passion is the fuel or key ingredient for the paths that are powerful enough to change our lives. Any path with passion begins when our dreams will no longer let us sleep. It starts with a longing in our soul that will not leave us alone. It is about our hearts getting out in front of our heads and putting feeling before thinking. It is fanning the flame that burns in us with actions that we may not have done before. Passion on any path is about the dedication, the hours, the sacrifice, the labor, being completely devoted to and compelled to do something. Passion is about doing now and asking questions later. It is the willingness to risk everything for one single moment, or the rest of our lives standing in the light of an idea, an idea that changed our lives for the better. It is being alive! **"Our passions are the winds that propel our vessel ..."(Proverbs)**

Sadly, there are the people living in the world without passion, without a burning desire to do anything. Or others that might have felt passion at one time but allowed it to die because of fear, or lack of faith, or self-worth. Fire spreads or burns out. Fire, if left alone will die; if we do not stoke the fire it goes away. Our passion can go away if we do not do something about it. Without passion, everyday looks and feels the same.

Passion, like everything else is overseen by the **Law of Cause and Effect**. We direct, but the Law provides. And here lies the difference between the passion in our lives and our everyday living without passion. If we look throughout history we will see almost miracle or magical moments where the universe bent over backwards to make someone's passion happen. We know by looking at the lives of people living their dreams that things fell into place because of their commitment to their dream. Sometimes that comes with hard work and struggle, but it also comes with some *inside* direction. **"The more intensely we feel about an idea or a goal, the more assuredly the idea, buried deep in our subconscious, will direct us along the path to its fulfillment."(Earl Nightingale)** Directing passion only seems bigger because our passions are not usually about everyday experiences. We do not feel the fire in ordinary living. But there is no big or little in the universe. The Law only knows direction pushed by thinking and feeling; it only knows to DO, without knowing what it is DOING. If we could put as much passion into loading the dishwasher as we did in going to the Prom or buying that first house, our everyday experiences would be more fulfilling and joyful.

168

It has been said that **Brother Lawrence of the Resurrection (born Nicolas Herman)** had to be chained to the floor to keep from levitating above the pots he would scrub. His great passion for God allowed him to experience God in what he called *common business*, no matter how mundane or routine. **"It is enough for me to pick up but a straw from the ground for the love of God."**(Brother Lawrence) Feeling the fire is the difference between the mundane and the sublime, and our whole lives could be filled with passion if we would only allow it.

We all have passion, whether we feel it or not. Prometheus only gave us what was rightfully ours, and that is why we are not punished for having it, but having it comes with responsibility. While we feel the fire, we can never let it go out, and so we have to fan the flames. We have to use the fire and not let it be **"blown out."**

Passion means greatness and we all have greatness in us, and greatness all about us. Passion is about the evolution of all human kind, the world, and the larger picture that we do not even see. It was the Passion of God that created the heavens, the stars, and us; so our passion whatever it is, is part of a bigger passion, **"As above, so below."**(Bible) If you do not feel you have passion in your life, look for it – it is there, **"Chase down your passion like it's the last bus of the night."**(Glade Byron Addams) If you feel passionate about something, you are on a path. **WALK IT!**

*"Passion, it lies in all of us, sleeping...
waiting... and though unwanted...
unbidden... it will stir... open its jaws and
howl. It speaks to us... guides us... passion
rules us all, and we obey. What other
choice do we have? Passion is the source of
our finest moments. The joy of love... the
clarity of hatred... and the ecstasy of grief.
It hurts sometimes more than we can bear.
If we could live without passion maybe we'd
know some kind of peace... but we would
be hollow... Empty rooms shuttered and
dank. Without passion we'd be truly dead."*

Joss Whedon

PASSION AND SPIRIT

Most "paths" are about our spiritual lives, deepening Spirit within us, healing our lives and the life of humanity, or making the world a better place. But passion can be about so many things, in so many ways, so the question here seems to be: **"Is passion about Spirit?"** We know about the passion of **Jesus, Buddha, John the Baptist, the Saints, St. Francis, Joan of Arc** and many others, but what about the rest of us: are our passions about God? In the long run, yes, I believe so. The place we arrive at when our passion has become an actual achievement is *fulfillment*. Fulfillment is the end product of purpose and meaning, passion is the vehicle to these ends. When our lives achieve fulfillment we help create a better world. How can anyone fulfill a passion and not see the magnificence that has unfolded in their lives and not wonder about that magnificence? How can anyone achieve something great and not have grown their faith, regardless of their belief? How can anyone witness the wonder of life and not see something behind the experience? Yes, eventually it all has to, in one way or another, come back to God. After all, **"All roads lead to Rome."(Proverbs)**

"When we feel passion for something, it is because we are remembering what it was that we came here to do. The more passion we feel, the more in alignment with Source we are, allowing this energy to pour through us with no hesitation. This is the way it was meant to be."

Karen Bishop

YOU ARE A SHINING CITY

You are a Shining City
Your light spreads from on high to the sea.
Your mansions and treetops gleam radiant;
Your fire burns eternally.

You are a Shining City
Your light beckons and draws near.
Your gates bright from the distance;
Your windows shiny and clear.

You are a Shining City
Your gardens glow in the dark.
Your fountains flow with lasting life;
Your radiance fills the ark.

You are a Shining City
Your truth is what we hold.
Your embrace raptures eternity;
Your love never grows cold.

You are a Shining City
Your light of life burns bright.
Your soul gives form to shadows;
Your heart stirs the dead of the night.

You are a Shining City
Your crown is for all to wear.
Your decree is "Let there be light;"
Your kingdom is for all to share.

You are a Shining City
Your arms forever unfurled.
Your dream cannot be hidden;
Your light is the light of the world.

"Here the angels raise the sun,
Like a veil being lifted in space.
A golden dawn inside the palace,
A smile on Creation's Face."

Chapter Fifteen: New Thought

THE NEW THOUGHT PATH

Many people walk the **New Thought Path**; they just do not know to call it that.There are many people in the world with positive belief systems, attitudes and minds, that use affirmations, believe good things for themselves and others; people who have faith in God and in themselves. A positive outlook is not limited to, nor does it belong to any one particular religion, spirituality, or faith. We all have our roots planted in the idea. If we go back and really look at **Ancient Wisdom**, we will see that a positive mind will create a positive life. But New Thought is about more than a positive mind; it is also about putting that mind into action in the individual life, and the life of the world.

New Thought is made up of organizations and individuals who share a set of metaphysical beliefs; and although New Thought uses the word "new," it really means "old," or ancient. The New Thought movement developed in the United States during the late 19th century. However, we can trace the basic ideas that make up New Thought back to the teachings of Jesus. New Thought teaches today what Jesus taught in the first century. He taught healing and faith – New Thought makes healing and faith the central principles of its theory and practice. He taught the power of belief – New Thought teaches the power of belief. He taught love and brotherhood – New Thought teaches love and brotherhood. He admonishes us to take no anxious thought for tomorrow – New Thought practices the Divine Supply. As with Jesus, New Thought overcomes sickness by health, evil by good, anger by love, error by truth. The things of God are positive, any negation is lack of God.

Some of the contributors who helped make up New Thought as we know it today, include Ralph Waldo Emerson, Henry David Thoreau, Thomas Troward, Emma Curtis Hopkins, Mary Baker Eddie, and Phineas Parkhurst Quimby. Founders of New Thought movements include, Charles and Myrtle Fillmore (Unity), Nona Brooks (Divine Science), and Ernest Holmes (Religious Science – Science of Mind). There are slight differences in theology and spiritual points of view, but their use of, and ideas about Universal Principles are the same.

Although New Thought was greatly influenced by Transcendentalism and some consider Emerson the father of New Thought, he did not put his theory to a definite concept that could be practiced. Phineas Parkhurst Quimby did and most consider him the father of New Thought. He believed that perfection is within everything and proved it in practice. He believed that a perfect man (human) stands behind all human trouble, sickness, and confusion. He believed that spiritual realization of this truth can and does heal. Thousands of people came to him to be healed. Recordings of his practice are published as The Quimby Manuscripts.

New Thought's basic mantra is, **"Change your thinking, change your life"** – even though thinking a new thought is simple, it is still very hard to do. Simple, because all we have to do is think another thought, but hard, because our negative thinking for the most part is habitual. Negative thinking is a way of life, and until we change it, we will revert back to it every time without fail. Positive thinking has to become a way of life and this is what New Thought is all about. New thought does not teach us what to think, it teaches us "*HOW*" to think. **William James** called it **"The Religion of Healthy – Mindedness."** We want a healthy mind, because a healthy mind will create a healthy life.

The key attraction of New Thought is "*personal power.*" Its theory is based on the fact that we are capable of manifesting remarkable changes in our lives. New Thought teaches us how to heal mind, body, and soul. It teaches us how to deal with our relationships, affairs, and yes, our checkbooks. Its theory is another attraction. There is no better subject than trying to answer the age old question, **"how do we live life?"** But I have found that too much theory and too little practice makes for a lot of good conversations, but few changes. The path to becoming healthy minded is all about *"practice."* New Thought gives us a container of tools to help us change our thinking and therefore our beliefs, and ultimately our lives. The tools are – prayer, affirmations, meditation, study, mindfulness, awareness, and yes, positive thinking. But we must practice them, not just talk about them. New Thought is a spiritual practice for the person who is ready to take responsibility for every area of their life.

The New Thought path is about a way of life that is closer to what God is, rather than what God is not. It is not looking at the world through rose colored glasses, but a positive view of life based on an understanding of our divinity as well as our humanity. It is being in the world but not of it in understanding that we are not our conditions, and any condition can be changed. It tells us that we are as God is (Spirit, Soul and Body) and that God is forever present in us, as us. It teaches us that we have dominion over our lives because we are made in the image and the likeness of God, and therefore creative by nature. This path is not about dogma, sin and sinners, hell or Satan, evil, or them and us. **"The New Thought path is about the recognition of the incarnation of the Spirit in everyone. The path is about believing in the unity of all life and manifestations of the One."**(Holmes)

At the heart of New Thought teaching is its profound understanding of **The LAW**, and our use of It to create better lives. It teaches that we do not have to beg God for what God has already given, but rather reveal those things in consciousness. It teaches us that everything is consciousness and that all of life happens through us and never to us. It is the path of the revelation of Truth and the Power of Mind. At the very core of New Thought stands the Power of prayer to heal; not only for ourselves, but others as well. One very rewarding path in New Thought is to become a license Practitioner or teacher. People who take this path spend three to five years taking classes and healing their own issues enough to be able to see the truth for others. Licenses are not given easily. There are many panels and exams, and Practitioners are held to a high standard of ethics.

My New Thought path is **Religious Science**; and I have found that the word science seems to trip-up a lot of people. For them, the word science can get Religious Science mixed up with other things. The word science its self can turn people off. I just keep it simple and say that what it teaches is provable. What is Religious Science? **Ernest Holmes** tells us, **"Religious Science is an educational as well as a religious movement and endeavors to coordinate the findings of science, religion, and philosophy, to find a common ground upon which true philosophic conclusions, spiritual intuitions, and mystic revelations may agree with the cold facts of science, thus producing fundamental conclusions, the denial of which is not conceivable to a rational mind."** Or more simply put, he also says, **"We are a group of people who believe in a truth, which we are endeavoring to prove. What is it? It is that God is all there is, not up in the sky but right here."**

My path has given me my life as I know it today. It has opened the doors to the other paths I have taken since first finding the book, the **Science of Mind**. It has afforded me many wonderful things and given my dreams form and experience. It has given me confidence, faith, and an unshakable understanding that we are far more than we seem to be. It tells me that the Kingdom of Heaven is here and now, and to experience it, I must become conscious of it. It gives me the freedom to see the golden threads of truth that run through every faith or religion. I can experience God anywhere, in a crowd or alone, in a temple, or a mosque or a synagogue. In New Thought there is no such thing as "your God <u>and</u> my God," there is only God, and God is found at the center of OUR being. This path is a threefold, it is a faith, a religion, and a way of life. For me this means a way of life that puts the knowledge of the Laws that govern my everyday life into positive use - a religion that I share with countless thousands of others who worship God in a beautiful way – and a faith that tells me that the life within me is Eternal and everlasting.

On the next page you will find the **"What We Believe"** statement written by Dr. Ernest Holmes in 1927. These statements reflect the core concepts of the Science of Mind teaching and contain within them Universal Spiritual Principles.

"What We Believe" By Ernest Holmes

We believe in God, the Living Spirit Almighty; one, indestructible,
absolute and self-existent Cause. This One manifests itself in and
through all creation but is not absorbed by its creation. The manifest
universe is the body of God; it is the logical and necessary outcome of
the infinite self-knowingness of God.

We believe in the incarnation of the Spirit in everyone and
that all people are incarnations of the One Spirit.

We believe in the eternality, the immortality,
and the continuity of the individual soul,
forever and ever expanding.

We believe that Heaven is within us and
that we experience it to the degree that
we become conscious of it.

We believe the ultimate goal of life to be a complete emancipation
from all discord of every nature,
and that this goal is sure to be attained by all.

We believe in the unity of all life,
and that the highest God and
the innermost God is one God.

We believe that God is personal to all who feel this Indwelling Presence.
We believe in the direct revelation of Truth
through the intuitive and spiritual nature of the individual,
and that any person may become a revealer of
Truth who lives in close contact with the indwelling God.

We believe that the Universal Spirit, which is God,
operates through a Universal Mind, which is the Law of God;
and that we are surrounded by this Creative Mind
which receives the direct impress of our thought and acts upon it.

We believe in the healing of the sick
through the power of this Mind.

We believe in the control of conditions
through the power of this Mind.

We believe in the eternal Goodness,
the eternal Loving-kindness,
and the eternal Givingness of Life to all.

We believe in our own soul, our own spirit,
and our own destiny;
for we understand that the life of all is God.

CONCLUSION

"When we rule our minds in a positive way, we choose the paths we want to walk down. We are no longer driven by egos. Instead, we become masters of our world and our destiny and create a life that is happy, joyous, and free."

John Marks Templeton

The many paths that I have taken on my journey have allowed me to be happy, joyous, and free. All the roads that I have traveled have given me something in one way or another. One of the paths that seems to serve me the best is inspiration. A book changed my life, and books continue to change my life and lead me in wonderful ways. For me, heaven is standing in front of my bookshelves, being guided by Spirit to the right inspiration for one of my talks, classes, or essays. The words of others inspire me to have words of my own. I trust that you have found words in this book that will inspire you to have words of your own. We all have something to say; we all have part of the answer. Sometimes we just need a little push. Many of the wonderful quotes in this book came from books such a **The Quotable Spirit** by **Peter Lorie & Manuela Dunn Mascetti** and from Internet sites that I find trustworthy such as, **BrainyQuote** and the **Quotegarden**. The quotes by **Ernest Holmes** can be found in the **Science of Mind** and **Living the Science of Mind** and many of his other books.

OTHER BOOKS BY DR. HOLMES:

This Thing Called You
This Thing Called Life
Words That Heal Today
Creative Ideas
The Art of Life
Effective Prayer
Ideas For Living
Richer Living
Thoughts Are Things
The Voice Celestial

MY LIST OF INSPIRATIONAL AUTHORS:

Aldous Huxley
Charles Fillmore
Deepak Chopra
Dr. Wayne Dyer
Eckhart Tolle
Eric Butterworth
Ernest Holmes
Gregg Braden
Howard Thurman
Joel S. Goldsmith
John Marks Templeton
Larry Dossey, M.D.
Matthew Fox
Neil Douglas-Klotz
Stephanie Soresen
Sri Aurobindo
Thich Nhat Hanh
Thomas Moore

AND POETS:

Hafiz
Tagore
Kabir
Kahlil Gibran
Rumi
Vivekananda
William Blake

MAY GOD BLESS YOU ON YOUR WAY.

JUST ON THE OTHER SIDE OF TOMORROW

Just on the other side of tomorrow,
There's a day brighter than today.
Only a dream's reach over my shoulder,
That's not so far away.

Here the angels raise the sun,
Like a veil being lifted in space.
A golden dawn inside the palace,
A smile on Creation's Face.

Just on the other side of tomorrow,
Life returns to be with the One.
Where the radiance of the full moon,
Shines in the light of the sun.

They linger there together,
Their voices sonnets of love.
They've called me out from my shadow,
To stand in the light from above.

Here every fallen flower,
Now in everlasting bloom,
And butterflies have forgotten,
Their fight inside the cocoon.

Here the winds make lullabies,
Rocks and rivers speak without sorrow.
My journey has led to rainbow's end,
Just on the other side of tomorrow.

Here there is an endless field,
Pure and gentle and fair.
A fire burns steady in every heart,
Souls soaring wherever they dare.

I too will soar as never before,
I'll fly between light's golden beams.
Above the dawn and inside the night,
I'll live my heavenly dreams.

Just on the other side of tomorrow,
My spirit steps out of me.
My soul is found and now unbound,
I live eternally.